Introduction to Coastline Development

Introduction to Coastline Development

EDITED BY

J. A. STEERS

THE M.I.T. PRESS

CAMBRIDGE, MASSACHUSETTS

Selection and editorial matter © J. A. Steers 1971

First published 1971 by Macmillan and Co Ltd

First M.I.T. Press Edition 1971

ISBN 0 262 19089 3 (hard cover)

Library of Congress catalog card no.: 75–138844

Printed in Great Britain

Contents

Acknowledgements

'Sea-level Changes during the Last 10,000 Years', by S. Jelgersma © Roy. Met. Soc., *Proc. Internat. Symposium on World Climates from 8000 B.C. to 0 B.C.* (1966)

'The Main Post-glacial Raised Shoreline of Western Scotland from the Firth of Lorne to Loch Broom', by S. B. McCann © Inst. Brit. Geogrs. 1966

'Trend Surface Mapping of Raised Shorelines', by S. B. McCann and R. J. Chorley © *Nature*, CXXV, 5 Aug 1967

'Marking Beach Materials for Tracing Experiments', by C. Kidson and A. P. Carr © *J. Hydraulics Division, Proc. Amer. Soc. Civ. Eng.* (1962)

'A Theory of the Development of Accumulation Forms in the Coastal Zone', by V. P. Zenkovitch © *Cahiers Océanographique*, XII, no. 3 (1960)

'The Relationship between Wave Incidence, Wind Direction and Beach Changes at Marsden Bay, County Durham', by Cuchlaine A. M. King © Inst. Brit. Geogrs. 1953

'The Influence of Rock Structures on Coastline and Cliff Development around Tintagel, North Cornwall', by Gilbert Wilson © *Proc. Geol. Assoc.*, LXIII (1952)

'The Form of Nantasket Beach', by Douglas W. Johnson and William G. Reed, Jr, *J. Geol.*, V (1910) © The University of Chicago Press 1910

'Methods of Correlating Cultural Remains with Stages of Coastal Development', by William G. McIntire © Office of Naval Research, Washington, D.C., 1959. This study is a by-production of a project sponsored by the Geographical Branch, Office of Naval Research, under contract Nonr 1575(03), NR 388002, with the Coastal

Studies Institute, Louisiana State University, Baton Rouge, Louisiana

'Coastal Research and its Economic Justification', by Per Bruun, *Congrès Internat. de Geog. Norden* (1960) © Royal Danish Geographical Society

Introduction

THE coast in this country has for many centuries been of interest to those 'that go down to the sea in ships, and occupy their business in great waters', but the scientific study of the coast is a very recent development. Even as a place for holidays the attraction of the coast is relatively new. The first coastal resort in this country – it was then called a spa – was Scarborough. In 1626 a certain Mrs Farrow noticed a medicinal spring in the cliffs, and made it known. She was aided by Dr Wittie, a good propagandist, who also advocated the drinking of sea water because, he claimed, it cured gout and 'all manner of worms'. The habit of drinking sea water for medicinal purposes continued even into Victorian times. But the attraction of the sea for bathing and for holidays gradually increased, and certain places became popular. The association of Brighton with the Prince Regent and of Weymouth with George III are well-known examples. But it was not until the building of railways and the increasing speed and ease of travel that the coast became popular in any general sense.

Even as late as the First World War scientific literature concerning our coast was scarce. There were of course several general books which referred to problems of erosion and accretion, but very few specific works. It is well worth while to glance through the bibliography of D. W. Johnson's *Shore Processes and Shoreline Development*, published in 1919. As far as this country is concerned books like Newell Arber's *The Coast Scenery of North Devon* (1911), curiously omitted from Johnson's list, and W. H. Wheeler's *The Sea Coast* stand out. There were indeed other books both in this country and elsewhere, but they were not numerous. On the other hand there were a large number of papers published in scientific journals, and with these must be included several publications, some in book form, dealing more specifically with harbour construction and related engineering problems. Of outstanding significance were the volumes of the Royal Commission on Coast Erosion (1907–11). The Minutes of Evidence, letters and other detail included in the early volumes are of great interest and should be read by any student of our coast in association with the final, summary volume. It is, I think, true that all these papers and books were of interest to relatively few people apart from those who, in some special or technical field, were concerned with the coast. This was to some extent the result of school and even university curricula. Geology and botany have long attracted distinguished scholars, but relatively seldom did they write on coastal phenomena. Geologists often wrote about the rocks in the cliffs and perhaps about cliff structures; botanists certainly collected and described coastal plants; but geomorphological and ecological studies are almost all comparatively new. Geography at that time was indeed much concerned with capes, bays and rivers, but it completely neglected the study of the ways in which such features are formed. Many schools, even as late as 1914, were primarily concerned with the teaching of classics and mathematics, history and languages and some science; other subjects were taught, but field work and field studies were almost unknown. Anyone who reads, for example, classic school stories of the early part of this century will realise how those 'other' subjects were regarded! That there were always exceptional schools, masters and pupils is fully recognised, but the present generation, with its great emphasis on field studies in the widest sense of the word, will scarcely, if at all, appreciate how much the attitude to nature study has changed even since the end of the Second World War.

There is, I think, little doubt that the better teaching of physical geography, both in schools and universities, has had an enormous influence on coastal studies. This is not the place to enlarge upon this theme, except to point out that in the last two or three decades geography in universities has to an increasing extent included within its scope the scientific study of physiography or geomorphology. Formerly this was regarded as much more a part of geology, and that it certainly is so is a fact, but in university curricula it is now usually much more fully treated in departments of geography than it used to be. Wherever it is studied I am convinced that in so far as coastal physiography is concerned it should be taught in such a way as to elicit the collaboration of specialists in other fields. Nowadays the inter-disciplinary nature of many subjects is recognised, and the student of coasts can usually seek and find the help of those in disciplines related to his own. This is as it should be, but there is in my view little doubt that the far better teaching of physical geography has led to the present much increased scientific interest in coastal matters. Let us not forget, however, the debt geographers owe to those geologists who were largely responsible for laying the foundations of physiographical teaching in universities.

Douglas Johnson's *Shore Processes and Shoreline Development*, already mentioned, represented a great step forward. For the first time the study of coasts was co-ordinated and put into a form which at once showed the possibilities of the subject. Until then the young student had had to be content with somewhat general accounts of coast phenomena in textbooks of geology or geography, and unless guided by a good teacher or possessed of great enthusiasm he was unlikely to pursue the subject any further. Johnson himself was a geologist, and in his book he was much concerned with processes, and the analysis of what is happening on a coast – the dynamic effects of waves and currents, of winds and tides. He was also concerned with the origin and evolution of coasts and of particular coastal features; he emphasised the significance of historical studies, and particularly in his later book, *The New England–Acadian Shoreline*, the importance of ecology in relation to coasts is shown to be of paramount interest. Moreover Johnson amassed a bibliography which comprised the work of specialists in other countries and, for the first time, gave us a good overall picture of the development of the study throughout the world.

Today, largely because of the motor-car, the coast is in great demand, not only in this country but in all with a seaboard in a reasonable climate. This pressure implies many problems, and many miles of coast, here and elsewhere, have been spoiled in one way or another. We are not directly concerned with planning, but it is relevant to call attention to the various bodies, national or local, which are concerned with conservation, that is with the right use of the coast. Certain stretches need complete protection; many others can be developed to some extent; some must be wholly given up to urban or industrial use. The National Trust, the Nature Conservancy, the Countryside (National Parks) Commission,[1] numerous county and local trusts and individual owners are all concerned with this problem. The recent very comprehensive regional and general reports of the National Parks Commission give a full analysis of how the coast is used in England and Wales. Scotland is less vulnerable, and so too is the coast of Ireland. But some other countries are under even greater pressure than England and Wales. We do not always appreciate how much of our coast is open to the public.

All this pressure has made an increasing number of people aware not only of problems of car parking and accommodation but also,

[1] The National Parks Commission was reincorporated as the Countryside Commission about the time the regional reports were in preparation. The final reports of the Commission should appear in 1970.

fortunately, of problems of erosion and accretion. It is of vital importance that more and more people should realise something of the natural processes at work on a coastline. It is a fact that on many coasts natural changes take place more quickly than in any other environment, apart from the sudden and catastrophic effects of volcanism or earthquakes or violent river floods. Anyone who takes notice can see the effects of a storm, of a change of wind and wave direction, of the effect of vegetation on mud flats, or of the continual recession of cliffs. What is more, any careful observer can add materially to our knowledge of coastal change. It is at least partly for this reason that an increasing amount of scientific observation and even investigation has taken place in recent years, but even up to the Second World War the volume of work in this country was not great. The Royal Commission's Report did not appear to lead to much active research. Engineers continued to build harbour works and sea defences, but for the most part on what might be called an *ad hoc* basis. Very little quantitative work had been attempted on the movements of beach and offshore materials, and the numerous references to raised beaches and submerged forests, although of considerable interest, did not give satisfying and comprehensive answers to the problems which the field investigations provoked. On the other hand Johnson's influence certainly made itself felt in America, Britain and other countries. Published accounts of parts of a coastline became far more analytical, and careful mapping of spits and other phenomena enabled us to appreciate how features like Dungeness or Cape Cod had evolved. A. G. Ogilvie's careful work on the coasts of the Moray Firth, published in 1923, deserved far greater attention than it received: it was the first attempt in this country to map and analyse a long tract of low coast.

We have been aware for many years of fluctuations of sea-level. That these fluctuations were closely related to raised beaches and submerged forests was fully appreciated, but it was not until 1915 that Daly focused attention on the matter in so far as it concerned coral reefs. It was realised as early as 1865 by Jamieson, in his classic work on Scotland, that there was a well-marked rhythm of coastal movement associated with the ice age. A tremendous amount of work has followed since then, but we are still unable to give a complete interpretation of events. This is partly because it is difficult to measure rise and fall of sea-level, relative to the land, accurately, and partly because it was not always clear precisely what was being measured. Tide gauges are now giving us a great deal of information, but since both land and sea are moving it is not by any means a simple problem to separate their effects. Moreover, exactly how does one measure, relative to the present level of the sea, the height of a raised beach? What part of the beach do you select for this purpose? Loose beach material, thrown up by waves, can give rise to very misleading ideas. A rock-cut bench is seldom if ever level; it usually slopes seawards, it may have a very irregular surface on account of the nature and structure of the rocks forming it, or it may be tilted as a result of land movements. A rapid perusal of many papers will also show that not only are precise measurements difficult to make, but also that in almost every locality there are far too few of them.

It is this lack of precision which detracts from the value of many researches on the coast. I do not in any way underrate what has been done; it was not until quite recently that accurate measurements became possible, and even now 'accuracy' must be regarded with caution, since few if any workers would claim that their conclusions are as precise as they would wish. In this country coastal research was spurred on by two serious floods – that which did so much damage in London in 1928, and the far more extensive storm surge of 1953. The former inspired investigation into flood levels in relation to tidal surges and also into the significance of a downward movement of the land in south-eastern England. The way in which this in its turn provoked a reconsideration of the interpretation of the successive geodetic levellings of this country in relation to movements of sea-level was of great interest to everyone interested in coasts. The 1953 flood, which affected a far greater stretch of coast both in these islands and on the Continent, made everyone realise something of the problems of defence against sea erosion and flooding. The work subsequently carried out, mainly by the Department of Geography in the University of Nottingham, on the replenishment of Lincolnshire beaches, almost totally displaced in the storm, has been of great significance. Since the Second World War there has been a great advance in the techniques of measurement of displacement of material along beaches and in the offshore zone. We are now in a position to evaluate, quantitatively and often with some exactness, just what is happening on a piece of coast. Previously we had to be content with qualitative assessments. Measurements of this nature must be carried out much more extensively before we can form an overall picture, but in many countries there is undoubtedly a much better appreciation of the problem than was possible even ten or fifteen years ago. Measurements of the vertical accretion of salt marshes have been made for some time and are comparatively easy to do. But we still lack detailed knowledge of cliff erosion. That recession is rapid in boulder-clay cliffs or cliffs of unconsolidated or unresistant materials is clear enough. A good many measurements have been made, for example in Holderness. But seldom has there been any planned scheme for recording loss over a long period of time, and no realiable knowledge is available on the effects of land water, draining out through the cliff face and so leading to slides, as compared with the effects of direct sea erosion. Other items of this nature could be given; the point is that there is great scope for all sorts of quantitative assessment on a coastline, and until we have these we must perforce be content with inadequate data for an understanding of coastal changes.

Allusion has already been made to the inter-disciplinary nature of coastal research. This, to many people, is one of its greatest attractions. The processes at work on a coast are complex, but their general effects are apparent to all observers. Wave action is the most potent factor, but its variations in intensity depend on many factors. Primarily it is governed by meteorological causes, and since these vary greatly in time and place it is difficult to make exact forecasts. Moreover in many latitudes wave action is very erratic in its impact. Tidal action is regular and predictable, but some of the most devastating storms – surges – follow from the interaction of tidal and meteorological effects, the particular incidence of which is difficult to foresee.

The nature of the coast itself is obviously important. The differences between the east and west coasts in England and Wales need no emphasis; in the United States there are profounder differences between the cliffed Pacific coast, the low-lying, flat and stoneless coast of the Gulf of Mexico, and the sandy low coast with its many barrier islands facing the Atlantic between southern Florida and New Jersey. Coasts of this type, and those formed by slightly resistant rocks, may change much in the course of time. Flat sandy coasts along which there may be barriers of sand or spits of shingle are very liable to rapid changes. Coasts where there are extensive deltas are subject to subsidence on account of the increasing weight of sediment brought down by the river. For these and similar reasons there is often a close association with archaeology and history. Remains of former human occupation are often found now well below sea-level, and changes in the growth and decline of some coastal towns are often closely related to changes in their harbours brought about by littoral drift. Not only delta coasts, but nearly all coasts, are subject to vertical movements relative to sea-level, caused by isostatic movements of the land masses – nearly always the after-effects of glaciation – and also by eustatic movements of sea-level. The former, like those in delta areas, are variable both in amount and in location; the latter are world-wide, but also mainly the result of the waxing or waning of existing ice sheets and glaciers.

Archaeological remains may allow us to measure the amount of vertical change, and often to date that change within certain limits. This is particularly the case if C^{14} methods can be applied to the finds. But change in a horizontal sense may also be profound. The growth of spits of sand and shingle, and the inundations caused by the blowing of coastal sands, afford numerous examples of the ways in which physiographical processes have played a role, often of great significance, in local or even national history. Historians and others would do well to recognise to the full the climatic changes that have taken place in this and other countries since, for example, the summit of the power of Rome. H. H. Lamb (*Geographical Journal*, Dec 1967) writes:

> The history of England, and that of Scotland, has generally been discussed without reference to change of weather or climate except as an occasional intruder whose appearances, pleasant or grim, have been treated as entirely haphazard. The suggestion that our climate has an intelligible history seems foreign, although in Scandinavia it cannot be ignored, and in Iceland the country's history consists of little else. Some claim to have 'proved' that the widespread social changes and unrest in the later Middle Ages were due to economic causes, and therefore not to any change of climate, as if such a thing would not be bound to show itself in the guise of economic stress or dislocation. The object of this study is to show, partly from the physical laws of the atmosphere's behaviour and partly from a wide range of direct observational evidence, that Britain's climate does have a history and that this is something which must concern us. It touches human affairs in many ways and the environment in which we live. And, because of the latest chapters in the story, it poses practical questions that we have to face today – especially in an age of long-term planning.

Anyone who has made a study of coasts will agree whole-heartedly with this! One need only refer to the inundations of sand in the thirteenth and fourteenth centuries in South Wales, to the final covering of the Culbin estate in 1694, to the effects of hurricanes in the Caribbean, the effects of storms on the coasts of the Low Countries and north-west Germany, the Delta plan in Holland and the proposed Thames barrage as illustrations of this theme.

The interrelation of physiography and botany is often extremely well illustrated on the coast. The way in which salt marshes develop as a result of silt and mud gradually accumulating on a sand flat, the gradual colonisation of plants, the upward growth of the marsh surface caused largely by the filtering effect of the vegetation on the silt-laden tidal waters, and the final passage to brackish-water marsh and even dry land, although this last stage is often the work of man. All visitors to a dune coast must be aware of the intricate relationship between sand supply, blown by the winds, and dune growth. The study of cliffs, still in its infancy, demands an understanding of geology, particularly if the structure of the cliffs is in question. The carving in resistant cliffs is often mainly controlled by joints and bedding, but it may be profoundly affected by igneous intrusions and the former history of the land mass in which the cliffs occur. We know next to nothing about the rate of cutting of rocky shore platforms and cliffs of resistant rock. How many of our cliffs are the sole work of the present agents working upon them? Some of the caves in the cliffs of Carboniferous Limestone in the Tenby peninsula in South Wales were formed long before the sea gained access to them; many of the cliffs of Cardigan Bay and elsewhere are but freshly exposed from their cover of boulder clay which spread over them thousands of years ago.

The civil engineer often has a profound influence on the coast. He is largely concerned with sea defences and harbour works, and in building these he has often had a considerable, and not always beneficial, effect on the adjacent parts of the coast. The positions, size and length of groynes and the nature of harbour entrances are two ways in which the littoral drift may be profoundly affected. It is often the case that the engineer has to act vigorously on some local piece of coast in order to prevent erosion. It is not always possible to investigate in advance how his solution will influence the coast on either side. Nowadays, however, problems of this sort are much more amenable to solution. The recent use of tracers – radioactive, fluorescent, or by means of specially marked pebbles – has enabled the engineer to experiment on the volume and direction of beach and offshore drift before building his defences. Even more effective is the use of scale models. The work of this type at the Hydraulics Research Station, Wallingford, has had an admirable influence on the possible effects of coastal defence schemes. It would be rash nowadays if the installation of any extensive groyne system were not preceded by a careful series of tests on a model. So many of the earlier schemes, however well intended, were erected without any experimentation and all too often only in relation to the locality immediately concerned.

It will be apparent that the study of coasts is not only of great interest from a scientific or a historical point of view, but that it is also of vital importance from the economic point of view. Reference was made on a previous page to conservation, and it was suggested that this meant the right use of the coast. At the present time it is difficult to foresee the future requirements of industry. No one is likely to object to the advantages of, for example, the gas supplies from the North Sea or of quicker travel to the Continent by means of hovercraft. But there has been no small outcry about how these and other undertakings affect the coast. Amenities and business enterprise are bound to oppose one another in some places, but it is no use being blind to the necessity and effects of new schemes as they affect the coast. There must be increasing liaison on a national scale between all who have an interest in the coast, and all sides must have a fair hearing. It is not easy to find solutions, but reasonable ones are much more likely to be attained if those who can see the coast as a whole, and can give time and thought to the problems in question, are able to take part in discussions from an early stage. More and more care is needed in the development of the coast. Coastal barrages across Morecambe Bay, part of the Solway, the Dee, even the Wash are all freely discussed. Advocates of certain attributes of such schemes make their points frequently and emphatically. The whole matter, in all its aspects, needs discussion on a national scale. A Morecambe Bay barrage may well benefit places like Grange-over-Sands by giving it a permanent water-level; its effect on west Cumberland, if a trunk road is built along the barrage, will be even greater. On the other hand will a Dee barrage necessarily improve the amenities of Anglesey? It would be very interesting to know, as a result of model experiments, what might happen if a barrage were built across the Wash.

Perhaps some readers of this book may at some time in their careers find themselves members of a council or committee that is concerned with the coast. It may be a river board, an urban or rural district council, a harbour authority or a national committee. It is vitally important that members of such bodies should, if possible, have some knowledge of the natural processes that take place on a coast. Problems of rateable value are usually of great local importance, and if taken alone may have a profound and perhaps undesirable effect on the coast. The more the local problem can be seen in terms of the national perspective, the more likely the right solution will be found.

To choose papers for a book of this sort is, to say the least, a rash undertaking, and one that exposes the chooser to an unlimited amount of criticism. There is, however, a companion volume to this – *Applied Coastal Geomorphology* – and although the two are distinct and can be used independently, the choice of papers for the two together is of greater significance than the choice for each separately. I have tried to do two things – first to cover the subject (though for obvious reasons not completely so), and secondly to do so in such a way that the papers should be representative of writers of several countries. But a severe limitation is necessarily placed on an editor, because it is obviously unwise to compile a book which will be too expensive for those for whom it is primarily intended. This means that the number of reprinted papers must be small and, what is perhaps even more serious, it is out of the question to include long papers or monographs. It would have been advantageous to reprint G. K. Gilbert's *The Topographic Features of Lake Shores*. Despite the limitations of its title this is a most valuable paper and every student of coasts should read it, since the analyses and descriptions given by Gilbert apply to features built in the former lakes Bonneville and Lahontan and now exposed so that they can be studied in three dimensions. I also regret that in neither volume has it been possible to include a comprehensive paper on the interrelations of vegetation and physiography in coastal areas. The reader will undoubtedly find other and perhaps more glaring omissions.

However, in this book emphasis has been laid primarily on matters which, although the papers may refer to a limited area, are of general rather than local or even regional value. Dr Jelgersma's paper is obviously of this nature. It gives an account of recent fluctuations in sea-level and is also a most useful summary of the different points of view, or hypotheses, dealing with this complex matter. So much remains to be found out on this subject that the wider the point of view adopted by the writer, the better. The student can easily follow the subject as it has been developed by other writers if he makes use of her admirable bibliography. Raised beaches are one of the phenomena touched upon in Dr Jelgersma's paper. Dr McCann's account is in space limited to part of the west coast of Scotland. But it is one of the first accounts that is based on a large number of detailed measurements. It is not the only paper of this sort; J. B. Sissons and his collaborators have made far more measurements in their work on the carse lands of the Firth of Forth. But since these measurements have largely been derived from shallow bores, the surfaces dealt with cannot be so well seen or appreciated as can those of the striking phenomena of western Scotland. Moreover McCann has not only taken infinite pains in measuring reliable heights, but his work has shown clearly how warping has played a significant part in the appearance of the beaches. Unfortunately we seem to be unable to get away from the terms '100-ft', '50-ft' and '25-ft' beaches. These, at best, mean very little and are often profoundly misleading. It is time for a revised nomenclature which will take account of the age and varying heights of particular raised shorelines in this country, especially in Scotland.

Since most modern work on sand and shingle beaches must involve the lateral displacement of material, it is essential to appreciate the methods used and the kind of results attained in achieving this object. Professor Kidson and Mr Carr have together collected a wide range of information and have both carried out a great deal of experimental work on various beaches in this country. Quantitative methods are by no means easy to use, and the student needs to be fully apprised of the practical difficulties involved as well as of the limitations of the various methods. It is noteworthy that methods of this sort can be, and are, used in model experiments. The first, and very successful, application in this country was made at the Hydraulics Research Station in a model built to investigate the silting in St Abb's Harbour, Scotland.

Because of the difficulty of language, by far the greater number of coastal physiographers in Western countries have been unfamiliar with Russian work on the subject. A few Russian authors have written papers – or the papers have been translated – in French or English. Recently the important book *Processes of Coastal Development* by V. P. Zenkovich has been translated, and an English edition published in this country. For the first time many are now able to read what has been done in the U.S.S.R. and also to find descriptions of coasts about which little if anything was known in the West. The paper that is included in this book gives in short form some of the views discussed more fully elsewhere, but it is a valuable summary which should serve as an introduction to Zenkovich's book, and to an appreciation of Russian work on coastal physiography in general.

Dr Cuchlaine King's paper on Marsden Bay is in one sense of very local interest, but must be read in its wider context. She gives an analysis of the changes which take place in the beach under the influence of varying winds and waves, and her work serves as a model for this sort of analysis in other places. Gilbert Wilson's account of the Cornish coast near Tintagel is also of limited interest in the sense that it deals only with a small stretch of coast. On the other hand it offers an excellent introduction to the way a geologist, interested in coastal forms, looks at and explains the features produced by marine action and in other ways on a coast of resistant rocks and complicated structure. Moreover it is a coast which shows to advantage Tertiary levellings and one in which the study of the evolution of its cliffs is by no means simple. Cliffs deserve much more study than they have yet received, and this paper shows clearly that their evolution in some localities has been intricate.

The paper of D. W. Johnson and W. G. Reed is now relatively old, but is included for at least three reasons. First, it is an early paper and is therefore of interest for that reason alone. Second, it is written in part by D. W. Johnson, who is one of the great figures in our subject. Third, it deals with a beach the changes in the evolution of which can be elucidated largely by means of the cliffs and related features in a cluster of submerged drumlins. To some extent also it indicates the methods employed by W. M. Davis – Johnson was a pupil of his. The student of the 1960s and 1970s will do well to ponder the writings of Johnson, and especially his well-known book.

W. G. McIntire's paper is of an entirely different type. It deals, as far as locality is concerned, with the Mississippi delta. This is an area not only of changing outline but also of subsidence, and McIntire's paper shows how cultural remains of American Indians can be used to interpret coastal evolution. It would have been easy to choose a paper dealing with the same subject in this country or on the adjacent shores of Europe, but this one has been deliberately chosen because it not only illustrates the principles involved, but does so in a place where archaeology and history have very different characteristics from those in north-western Europe. The final paper, by Dr Per Bruun, needs no introduction. Per Bruun is a Dane and did much of his earlier work in Denmark. For several years he has lived in America, and is an authority, amongst other things, on the coast of Florida. It is an excellent summary of the economic reasons for coastal research. Even now we are not willing to spend as much on this type of research as we ought. We are far too apt to wait until catastrophe occurs and then do our best to alleviate it. In this country the research provoked by the floods of 1953 was certainly considerable. I do not refer so much to academic studies, but to the research on the building and maintenance of sea walls, usually composed of clay and sand, along the lower parts of our rivers and estuaries, to the effects of vegetation on such walls, to the effects of groynes and other sea defences, and to the movement of beach material and other phenomena. Even now we have not seriously undertaken research on cliff erosion. Per Bruun shows only too clearly that money spent on research may later save great expenditure and even loss of life and property.

It would be easy to add to the number of papers in this book in such a way that more writers were represented and more examples taken. Some aspects of coasts have been omitted, but that is inevitable in a work of limited size. The primary object has been to draw attention to certain important matters that must be taken into consideration by anyone interested in coastal phenomena. It is not always easy to obtain access to a large number of journals, and even if a particular journal is available, it is seldom that there is more than one copy. This book and its companion volume, whether they are used singly or in combination, should enable the student easily to expand his reading in such a way that he can have at hand examples of the writing of people who have devoted much of their time and energy to coastal studies. It is valuable to have these papers in a handy form, since not only can much be learnt from each one, but a comparison of the styles of the several writers and the way in which they develop their subjects can be made which may well be of help to anyone wishing to attempt research in the subject. All of the papers are accompanied by bibliographies or by notes, and these will help the reader to extend his range. One of the papers – that of Zenkovich – has been translated; nowadays, fortunately, it is not too difficult to obtain translations of papers. Nevertheless it is a great advantage to read papers in the language in which they are written, but this is an asset not enjoyed by everyone. The omission of papers by French, Italian, Spanish or other authors is regretted, but space and cost imposed rigid limitations.

Coastal research is being actively pursued in many countries. The Commission on Coastal Geomorphology of the International Geographical Union has helped greatly in making known to one another those pursuing coastal research in different lands. It is of real value to meet and discuss with experts, and one valuable function of international meetings is that it allows the student to do this, and also to have the opportunity of seeing a part of the coast of another land under expert guidance. The chance of seeing more of our own or any other coast should never be lost. No two places, even a mile apart, are subject to quite the same influences. Even on a straight coast wave action is variable and differs somewhat from one part to another. Moreover it is essential, if possible, to see the coast under different conditions. The effects of even a minor storm may completely alter the configuration of a beach, and there is an enormous disparity between what happens at high and low water, especially on a coast where the tidal range is appreciable. Coastal research offers wide possibilities, and changes, often of considerable magnitude, can and do take place in a relatively short time, a fact which undoubtedly makes the work more attractive to many workers. There should be more than enough in this volume to urge the reader into the field to do some work of his own!

In conclusion it is worth emphasising the great changes that have taken place in the study of physiography since the dominance of W. M. Davis and his immediate successors. Davis had a profound influence, particularly in America but also in Europe. Today his writings are perhaps of more interest to the historian of physiography than to its practitioners. His explanatory descriptions of landforms threw a new light on the evolution of landscape, but they were based on rigid rules of deduction. He introduced numerous terms and definitions, and almost created a new language. R. J. Russell rightly says in his *River Plains and Sea Coasts* (1967) that his 'elegance in logic commonly outweighed the presentation of sound field observations'. To a limited extent this is true of some of D. W. Johnson's coastal writings. Nevertheless the student should neglect neither the work of Davis nor that of Johnson; he should in fact study those parts relevant to the subject-matter of this book and note how and why attitudes have changed. W. M. Davis's *The Coral Reef Problem*, never an easy book to read, is nevertheless worth comparison with recent work on coral reefs and islands; D. W. Johnson's *Shore Processes and Shoreline Development* should be studied again after reading recent works on the same subject. The earlier work should in no sense be underrated; the student will learn much from a consideration of the changing points of view.

Any observer today will soon find that coastal problems can seldom, if ever, be fully solved by Davisian methods. The modern student, with his knowledge of statistical and quantitative methods and new techniques, a more detailed understanding of the relative movements of land and sea, a much greater acquaintance, gained through ease of travel, of a wide range of landscapes, and an awareness that tomorrow may well bring forth some new information which will cause him to re-think his problem, realises that there is no simple answer to his research. He must observe and collect as many facts as possible and try to interpret them in the light of experience – his own and that of others. Then he may begin to understand how and why change takes place on a coastline.

1 Sea-level Changes during the Last 10,000 Years

S. JELGERSMA

SUMMARY

In this paper the results of various studies of sea-level changes are evaluated. The main problem in all studies is the possibility that tectonic movements have influenced observed levels. In consequence, opinions on eustatic sea-level changes during the last 10,000 years are rather divergent. Some agreement has been reached on the movements of early Holocene sea-level, but the changes in level during the last 5,000 years are much disputed. Three different opinions on the movements of eustatic sea-level are held: an oscillating, a steady and a rising sea-level. At the present time the rising sea-level concept seems the most likely. Evidences for high Post-glacial sea-level stages are questionable.

The Holocene sequence in the Netherlands and adjacent areas has been treated and compared with dated sections from the Gulf Coast. The opinion is expressed that the cyclicity in Holocene coastal deposits shows the effect of a succession of wet and dry climatic conditions.

1. INTRODUCTION

AN oscillating sea-level during the Pleistocene was first recognised in 1842 by Maclaren, who introduced the theory of glacial control. This means that sea-level changes are due to changes in climate. It is a generally accepted fact that changes in temperature disturb the state of equilibrium between water in the ocean basins, moisture in the atmosphere and the water precipitated on the land that flows through rivers back to the sea.

So, if the warm interglacial climate turns into a cold glacial climate, the precipitation in large areas changes from rain into snow. The latter does not return back to the water of the ocean but stays on the land locked up in ice sheets. This results in a lowered sea-level. It has long been accepted that this mechanism, which we call glacial control, has been very important during the Pleistocene. The extreme changes in climate characteristic of this period have involved extreme changes in sea-level.

It must be stressed that any change in climate that alters the

amount of ice on the land will affect the level of the oceans correspondingly. These changes in sea-level are called eustatic.

In this paper the changes in sea-level during the last 10,000 years
(the Post-glacial or Holocene) will be discussed. It must be realised
that during this period the amount of rise of sea-level has been much
smaller than during the preceding Late-glacial, and so we only deal
in this paper with the final stage of the restoration of ocean-level.
For this period radiocarbon dating has been very useful, as it brings
into focus the time element in the studied shorelines. The use of
this method, by which carbon-containing material such as shells and
peat can be dated, has been of great benefit to studies of sea-level
changes. Before its development only relative age determinations
such as provided by pollen analysis could be used for the construction
of a timetable of former sea-levels.

During recent years numerous new investigations on sea-level
changes based on radiocarbon datings have given detailed curves of
the Holocene sea-level. The quantity of observations on former Holocene sea-levels might lead us to expect an unambiguous interpretation,
but on scrutinising the available results the clear picture vanishes.

A certain agreement is reached on the movements of the early
Holocene sea-level, but the changes in level during the last 6,000
years are much disputed.

The controversial opinions on this subject may be divided into
three groups, which could nearly be called three parties, each
supported by dedicated followers. The first group claims to have
evidence that sea-level has been rising rapidly until the end of the
Atlantic to about 3 m. above the present level, and fluctuated after
that time with an amplitude of 6 m. (Fairbirdge, 1961). This is the
theory of an oscillating sea-level after 6000 B.P. The second group
favours a steadily rising sea-level during the Holocene, reaching its
present level at about 3600 to 5000 B.P. This is the theory of a
standing sea-level after 3600 B.P. (Godwin *et al.*, 1958; Fisk, 1951;
McFarlan, 1961).

The third group denies any sea-level higher than the present during
the Holocene and also a standing sea-level after 3600 B.P. The
Holocene rise in sea-level is seen as a continuous one, diminishing
with time but going on until the present day (Shepard, 1960, 1963).

These different opinions on the movements of eustatic sea-level
during the Holocene are due to the fact that a study of this subject
only gives evidence of changes in level in a restricted area. Changes
in level may be attributed to three factors: eustatic movements of

sea-level, tectonic movements and compaction of sediments. We have already dealt with eustatic movements of sea-level and their world-wide and similar effect. Tectonic movements may have an important share in changes in level, particularly in mobile belts such as the circum-pacific mountain ranges where earthquakes and volcanoes are active. Uplift and downwarping may occur in these areas. Tectonic movements are also important in areas recovering from isostatic downwarping due to ice-loading. The latter is evident from the elevation of marine terraces (the so-called raised beaches) in areas which were covered by an extensive ice-sheet during the last glaciation. Studies of raised beaches in Scandinavia, Alaska and Canada indicate them to be caused by a dome-like uplift. In the centre of the uplift early Holocene shorelines may be found at an elevation of some 225 m. above sea-level. Many deltaic areas, such as the Mississippi delta, indicate tectonic downwarping as illustrated by cross-sections through Tertiary and Pleistocene deposits. The same situation may be found in the North Sea basin.

Compaction in sediments is caused by load, consolidation as a function of time, and by changes in hydrological conditions such as drainage. The compaction is highly dependent on the type of sediments and consequently may vary locally. In sandy sediments compaction is relatively low, in clays higher, and in peat very high.

Compaction may be a very important contributor to change in level in those areas where the Post-glacial rise of sea-level has caused the deposition of 10 to 20 m. of clay and peaty sediments. I think of the coastal plains of the Netherlands, Germany and the Gulf Coast of the U.S.A.

If we look upon the three discussed factors it is obvious that eustatic sea-level changes are the same the world over, but owing to tectonic movement and compaction this constancy is obscured. This is the reason why the results of studies on eustatic sea-level changes are so controversial.

We may avoid the factor of compaction to a certain degree, by taking as evidence of former sea-level only deposits resting on sediments, which show little or no compaction (Jelgersma, 1961). But the question remains whether it is possible to separate tectonic movements and eustatic sea-level changes.

If we want to study the eustatic sea-level changes during the Post-glacial we should not investigate an area recovering from isostatic downwarping like Scandinavia or an area where earthquakes or fault systems are important. We should prefer the so-called stable areas. Even then the results are questionable, since it is hard to believe in the stability of any area in the world.

Accordingly we may never be able to construct a good curve of the eustatic sea-level changes during the Post-glacial. This is, however, a bit too pessimistic, for if all over the world many careful studies on fossil shorelines dated by C^{14} could be made, an approximation to the movements of Post-glacial eustatic sea-level should be possible.

Special attention should also be paid to the ecology of the material used for dating. Only material from environments of deposition where relationship to sea-level is established should be used to construct former sea-levels.

When using peats we need to know how the ground-water table by which the peat has been formed stood in relation to the sea-level. This is, however, in many cases detabable (Jelgersma, 1961) and can be a serious source of error. Another source of error in using peats is that contamination by older material makes the sample seem too old when dated by radiocarbon; penetration by roots, however, makes it appear too young. Careful inspection of all peat samples to be dated is necessary. Contamination, however, is not always noticeable and the obtained age should be checked by other methods of investigation, e.g. pollen analyses.

The dating of shells by means of radiocarbon may also lead to a wrong age determination. The possible exchange of carbon ions subsequent to deposition of the shells may result in erroneous determinations of carbonates of different ages. An enrichment or differential loss in carbonate is likely to result in a non-representative age determination. Another source of error when dating shells is due to the fact that shell beds may consist partly of shells eroded from older deposits and washed together with fresh shells. The dating of bivalves and shells in living positions in the investigated sediment is preferred.

From the above-mentioned sources of error the conclusion may be drawn that even if in a certain area a series of radiocarbon dates of former sea-levels is available, the drawing of a curve of the changes of level is quite often not an easy procedure.

2. EVIDENCES FOR A POST-GLACIAL HIGH SEA-LEVEL

As mentioned in section 1, a group of students on Holocene sea-level changes has evidence that points to a higher sea-level than the present one during the last 6,000 years. The most controversial point of Post-glacial sea-level is that according to the theory of glacial control, during the Post-glacial thermal maximum, the Hypsithermal, higher sea-levels than the present would be expected. This high sea-level has been adovcated by Daly, who found at several places in the Indian and Pacific Oceans remarkably fresh-looking coral terraces elevated about 6 m. above the present mean sea-level. As the corals only grow up to the low-tide level, he inferred a world-wide sinking of ocean-level in recent time (Daly, 1920, 1934). He supported the idea that at the end of the Hypsithermal or Atlantic, sea-level was about 6 m. higher than at present, and that during the succeeding cooler sub-Boreal period sea-level dropped to its present level. This whole process would agree well with the idea of glacial control.

The question arises, however, as to how quickly the sea-level will react to these small climatic fluctuations. If the climate changes from warm to relatively cool, the fauna will quickly respond to this cooling-off and so will the flora, but with some time-lag caused by the immigration of plants. As far as sea-level is concerned there might be a greater time-lag involved caused by the growth and shrinkage of glaciers. Another point is that even if the climate cools off, the temperature may still be high enough to continue the process of melting of the remnants of the ice sheets. It is known from studies of the last interglacial, the Eemian, that sea-level was still rising considerably after the so-called climatic optimum of the Eemian (Zagwijn, 1961; West and Sparks, 1960).

The same may have occurred during the Holocene. Accordingly the transition from the Atlantic to the sub-Boreal period is not necessarily marked by a drop in sea-level.

During recent years studies have been restarted in the type locality of the Post-glacial high sea-level, the elevated coral terraces in the Pacific. Shepard (1963) summarises the preliminary results of these studies as follows:

However, more careful study of many of the atolls in the Pacific (see for example Emery *et al.*, 1954, and McKee, 1959) showed that the elevation of these islands was based almost entirely on the piling-up of rubble by large waves attacking the reef front and breaking off blocks. There are undoubtedly elevated reefs and wave-cut benches in the coral islands as well as in volcanic islands like Hawaii, but it should be borne in mind that these elevated reefs and benches could be due to warping or to an interstadial high stand.

The last-mentioned hypothesis agrees well with the results of the radiocarbon dates of shells from the 5-ft and 12-ft terraces along northern Oahu, which point to a Pleistocene age (Shepard, 1963, 1964). Much more work has to be done on this subject, especially a careful selection of the material from the terraces used for dating and the application of different methods of age determinations like Th^{230}. Absolute age determinations by means of Th^{230} growth method and C^{14} on samples of coral from borings on Eniwetok Atoll suggest no higher Holocene sea-levels than the present one (Thurber *et al.*, 1965; Broecker and Thurber, 1965). The data available point to a slowly rising sea-level after 6000 B.P. Before final conclusions can be drawn, more certainty has to be obtained about tectonic movements in this region. More dated samples of coral from several series of borings through an atoll will give interesting evidences of sea-level changes during the Holocene and the Pleistocene period. Owing to the above-mentioned facts, the type locality for Post-glacial sea-levels, the elevated coral-reef terraces, might not be repesentative, as some are of Pleistocene rather than Holocene age.

Other shorelines, like the Silver Bluff, along the south-eastern United States Atlantic coast, elevated about 10 ft above sea-level, are no longer believed to represent a Post-glacial high sea-level as stated by MacNeil (1950). Radiocarbon dates suggest a Pleistocene age (Hoyt *et al.*, 1964, 1965).

Elevated shorelines (10 ft) of the Bahama islands are, according to absolute age determination, also reported to be of Pleistocene age (Broecker, 1965).

Another important area where high Post-glacial sea-level stages are reported is New Zealand and Australia (Fairbridge, 1961; Gill, 1961, 1965; Schofield, 1960, 1964; Ward, 1965). Many elevated shorelines dated by C^{14} which are thought to represent the eustatic fluctuation of sea-level during the Holocene may be found in these regions. The report of the ANZAAS Quaternary Shorelines Committee of the Hobart Congress (1965), however, mentions that the sea-level curve of New Zealand by Schofield is strongly influenced by tectonics. Accordingly the Post-glacial heights of sea-level could be caused by warping.

The data of the curve of Schofield (1964) are obtained from a study of beach ridges and intertidal flats in a chenier plain in the Firth of Thames, New Zealand. The curve presented in Fig. 1.1 shows sea-level of 4000 B.P. at about 2 m. above the present level. After that time sea-level dropped with minor fluctuation to the present level. Another interesting point in Fig. 1.1 is the curve of the calculated sea-level. This curve, claimed by Schofield to be the eustatic one, has been computed from the dated raised strandlines of Fennoscandia.

This method used by Schofield (the calculations are made by Thompson) is of interest. The results, however, are questionable because of the uncertainties associated with establishing isostatic recovery curves.

In Australia the eustatic sea-level changes during the Holocene are a much disputed subject. The latest data mentioned by Hails (1965) do not point to the high Post-glacial stages as reported by Fairbridge (1961), Gill (1965) and Ward (1965). Apparently New South Wales, including the Macleay deltaic plain, has given no evidence for the high sea-level concept. In south-east Australia, however, emerged shell beds dated by C^{14} suggest higher sea-level than the present one during the Holocene (Gill, 1965).

High Post-glacial sea-level stages deduced from investigations in North Africa (Tunisia, Algeria and Morocco) and other parts of the Mediterranean are claimed by Fairbridge (1961) to be of eustatic origin. These shorelines are, however, situated in a mobile belt (Gourinard, 1958), so the question arises whether they are caused by eustatic movements. A tectonic origin seems more likely.

If we look at the areas where high Post-glacial sea-level stages are reported, the occurrence of mountain ranges quite near the coast is a striking phenomenon (North Africa, New Zealand and some parts of Australia). On the other hand studies that fail to support the high Post-glacial sea-level concept have all been carried out in flat areas, quite often deltaic areas, without any important mountain ranges in the vicinity (Netherlands, north-west Germany, the Gulf Coast of the United States and Florida). This is nicely demonstrated by the sea-level curve of Fairbridge (see Fig. 1.2). Dated shorelines from all over the world are used to construct the curve of the oscillating sea level during the Holocene. Fairbridge's low sea-level stages are based on the presence of peat deposits in the flat areas without any important mountain ranges. The high stages, however, are inferred from dated shorelines in areas where mountains are present near the coast.

Summarising the results of the investigations on Post-glacial sea-level stages, it can be stated that they suggest, but certainly do not prove, higher sea-level than the present one.

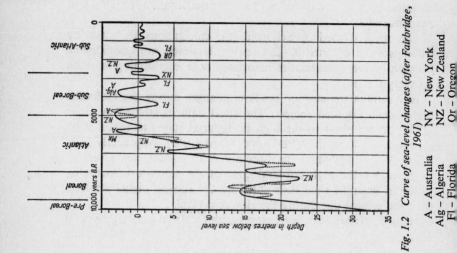

Fig. 1.2   Curve of sea-level changes (after Fairbridge, 1961)

A – Australia   NY – New York
Alg – Algeria   NZ – New Zealand
Fl – Florida   Or – Oregon

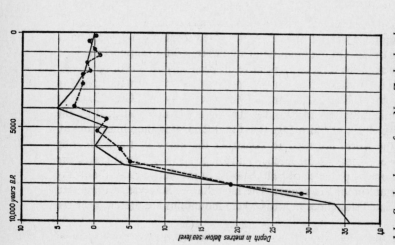

Fig. 1.1   Sea-level curve from New Zealand and a general eustatic curve calculated from raised strandlines in Sweden and Norway (after Schofield, 1964)

Calculated eustatic curve

3. THE STEADY SEA-LEVEL CONCEPT

The hypothesis of steady sea-level after 5000–3000 B.P. originated in extensive studies on the Holocene deposits of the Gulf Coast region of the U.S.A. Fisk (1951), Gould and McFarlan (1959), Le Blanc and Bernard (1954), McFarlan (1961) and Coleman and Smith (1964) have published details of changes in level on the Gulf Coast based on the geomorphology of the coastal plain and the stratigraphy of the Holocene supported by radiocarbon datings. These authors distinguish in the changes of sea-level after the last glaciation a rising and a steady sea-level stage. The latter should have occurred during the last 5,000 years. This is concluded from the lack of difference in elevations between the older and the younger beach accretions along the Gulf Coast, which suggest no sea-level changes during the last few thousand years. Owing to the fact, however, that the dating of the older beach accretions is based on radiocarbon datings of shells from the beaches, a possible source of error is introduced as many of the shells may be older than the beaches. So there is a possibility that the beaches are younger than indicated by the datings.

Further investigations by Gould and McFarlan (1959) and McFarlan (1961), supported by dated samples from deposits in the chenier plain of south-west Louisiana, have given indications that sea-level reached its present height 3,000 years ago instead of 5,000 years ago. Coleman and Smith (1964) have made subseqeuent investigations on the coastal marshes east of the chenier plain. The obtained data, not corrected for supposed tectonic subsidence and compaction in the underlying sediments, point to a slow relative rise in sea-level during the last 3,000 years.

The authors, however, introduce a correction factor for the supposed tectonic subsidence of 7 cm. per 100 years. This correction results in a stationary sea-level during the last 3,600 years. Scholl and Stuiver (1965) have given an excellent discussion of the study of Coleman and Smith. The applied correction factor of 7 cm. per 100 years is believed by Scholl and Stuiver to be much too high, as the chenier plain is thought to be a rather stable area (Gould and McFarlan, 1959). If the correction is reduced to a rate of tectonic subsidence of $3\frac{1}{2}$ cm. per 100 years, the relative sea-level curve is nearly identical with that derived from investigations in south Florida (Scholl and Stuiver, 1965) and on the Atlantic coast. Discussing results from sea-level investigations in south Florida, Scholl

ICD B

and Stuiver state: 'the carefully considered and collected Louisiana submergence data can be interpreted to indicate a slight (about 1·6-metre) rise in sea-level during the last 3,600 years. Therefore both Louisiana and Florida submergence data suggest that a strict interpretation of the stable sea-level hypothesis should probably be abandoned.'

The graph of eustatic sea-level changes published by Godwin, Suggate and Willis (1958) supports the steady sea-level concept, i.e.

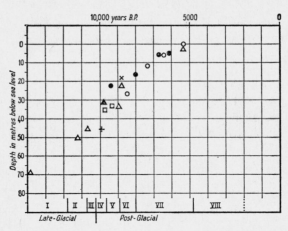

Fig. 1.3   Graph of eustatic sea-level changes (after Godwin, Suggate and Willis, 1958)

△ Gulf of Mexico   ☐ Southern Baltic
● Christchurch, New Zealand   ○ British Coast
+ Foxton, New Zealand   ▲ F.A.O. Persian Gulf
× Melbourne, Australia

after 5000 B.P. (Fig. 1.3). The curve is based on radiocarbon datings from deposits formed close to their contemporary ocean level.

The samples are from all over the world, including the British coast. The data of the standing sea-level after 5000 B.P. are obtained from England and the U.S.A. Gulf Coast. The evidence of the last-mentioned area is discussed above, the steady sea-level after 3650 B.P. instead of after 5000 B.P. being championed by several authors but denied by others. The samples from the British Isles have been collected in the Somerset region, which is, according to Godwin, tectonically stable. The evidence of stable sea-level is derived from the base of the Upper Peat, on top of marine deposits in Tealham Moor, Somerset, which is lying at the present sea-level. According to radiocarbon datings the base of the peat has an age of about 5400 B.P. and it is concluded at that time sea-level stood at its present height. The assumption that the underlying marine deposit was alluviated up to the mean sea-level of that time is questionable (Jelgersma, 1961). An alluviation between mean sea-level and high tide seems more likely.

If we look at the evidence that should support the steady sea-level concept, some doubt arises of its reliability.

4. THE CONTINUOUSLY RISING SEA-LEVEL CONCEPT

A continuous rise in sea-level during the Holocene, diminishing in time but going on until the present day, was first recognised by Shepard and defended by him in several articles (Shepard and Suess, 1956; Shepard, 1960, 1963 and 1964; Shepard and Curray, 1965). In his 1963 and 1964 publications he presents a time–depth diagram of C^{14}-dated shells and peat samples from all over the world. The material used for dating is believed to have been formed close to its contemporary sea-level. This includes the investigations on sea-level changes on the east coast of the U.S.A. These investigations, some details of which are published by Bloom and Stuiver (1963) and Redfield and Rubin (1962), point to a continuously rising sea-level during the Holocene.

The graph presented by Shepard (see Fig. 1.4) is suggestive because the majority of the dates indicate no higher sea-level than the present one during the Holocene and also a sea-level slowly rising after 4000 B.P. The data pointing to higher sea-levels than the present one are from regions like Australia, as discussed previously in this paper.

Investigations made by Scholl (1963 and 1965) and Scholl and Stuiver (1965) in south Florida have given evidence that approximately 4,400 years ago sea-level was about 4 m. below its present height. In the next thousand years the sea rose rather quickly to a level of about 1·6 m. below that of today. Accordingly the rise in sea-level during the last 3,500 years is small (1·6 m.) but noticeable (see Fig. 1.5).

The data used are derived from the age of the base of the mangrove swamp, which overlies fresh water deposits. The mangrove swamp has been spreading landwards during the rise of the Holocene sea-level. In some areas the age of the former sea-level has been determined by dating marine shells. Scholl believes that during the Post-glacial south Florida was probably a tectonically stable region, and considers

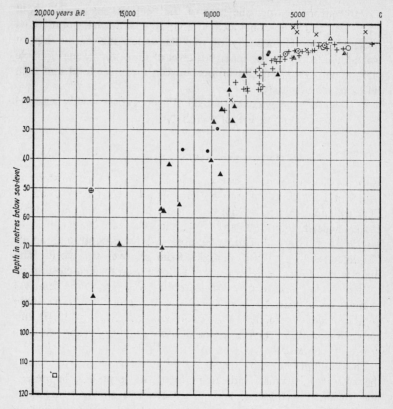

Fig. 1.4   The rise of sea-level indicated by C¹⁴ dates of organisms and plants that lived close to sea-level in relatively stable areas (after Shepard, 1963)

▲ *Texas shelf*   O *Florida*
+ *Holland*   △ *Ceylon*
● *Eastern Argentina*   □ *Western Mexico, off broad coastal plain*
◉ *South-west Louisiana*   ⊕ *Western Louisiana shelf*
✕ *Australia*

his sea-level curve to represent the eustatic one. The sea-level changes during the last 4,400 years are indicated by Scholl with a smooth curve. His data, however, give rise to the drawing of a curve showing fluctuations in the rise of sea-level. At the moment these fluctuations are not considered to be real because of the sources of error, and more data are necessary before final conclusions can be drawn. In Scholl's work it is clear also that because of the tectonic

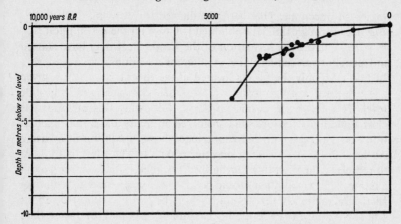

Fig. 1.5   Curve of sea-level changes in south Florida (after Scholl and Stuiver, 1965)

stability of south Florida his curve could be explained by eustatic movements. The question still arises about the part tectonic movements could have played in the measured changes in level.

The coastal plain of the Netherlands has been subject to several studies in sea-level changes (Bennema, 1954; van Straaten, 1954; Jelgersma, 1961). In the last-mentioned study the evidence for changes in level is based on C^{14} datings of the peat layer directly overlying the inclined Pleistocene surface at various depths and places in the coastal region of the Netherlands. As the peat layers (the so-called Lower Peat) are considered to be formed by rising ground water controlled by rising sea-level, the datings may serve to establish a curve of sea-level changes in this region.

The dating of the base of the peat layers resting directly on Pleistocene sand has an advantage and a disadvantage. By using this method the compaction in underlying sediments can be eliminated, as the Pleistocene sands show nearly no compaction. The disadvantage is the questionable reliability of the datings as indicators of sea-level changes. The assumption is made that the Lower Peat was formed in close relation to sea-level, but in places where upwelling water from the hinterland or stagnant ground water in local topographic depressions is present, this may not be the case. A serious source of error may be present in the data obtained (Jelgersma, 1916). For this reason the curve representing the movements of sea-level in the coastal plain of the Netherlands is drawn with certain restrictions and the reliability of the points under consideration is taken into account. The method followed for drawing a curve (see Fig. 1.6) through the data obtained is based on the assumption that in several places the formation of the Lower Peat may have started independently of sea-level (i.e. at different elevations above sea-level). In that case only the points located in the lowest places for a given age represent a ground-water table which coincides with high-tide level. This results in a smooth curve. Following this method it is clear that all aberrations are considered as errors and not as fluctuations of sea-level. The curve obtained shows a continuously rising sea-level gradually levelling off after 6000 B.P. The curve is thought to represent the high tide during the Holocene, as peat in the humid climate of a coastal plain should be formed coincident with high tide. In the Netherlands this formation will be near the coast at about 1 m. above mean sea-level; farther inland this may be lower than the mentioned average. Accordingly the dotted line is supposed to represent the movements of the mean sea-level during the Holocene.

For the full treatment of the presented data, especially for those of the province of Zeeland, which are slightly divergent, reference is made to Jelgersma (1961). The curve presented in Fig. 1.6 is slightly different from Jelgersma (1961), since the dates needed corrections for the Suess effect.

As in all discussed curves tectonic movements must also be included in the sea-level curve of the Netherlands. In the Netherlands a slight tectonic downwarping due to its position in the North Sea basin must be expected. Several authors have tried to calculate this downwarping from the elevation and age of the interglacial Eemian marine deposits (Tesch, 1947; Bennema, 1954; Jelgersma, 1961). The tectonic subsidence values obtained by this method are small, between $1\frac{1}{2}$ and $3\frac{1}{2}$ cm. per century. Accordingly tectonic subsidence during the Holocene is thought to be not very important and the presented sea-level curve should for the greater part be due to eustatic rise in sea-level. Tectonic movements, however, are not a gradual and continuous process at all and may vary strongly in time, so that the subsidence data are questionable, and consequently the sea-level curve.

A sea-level curve for the Netherlands and the adjacent North Sea basin is given in Fig. 1.7. The data for the early Holocene sea-level changes are derived from pollen analyses of moorlogs dredged off the Dutch coast. The curve indicates a sea-level stage of about 37 m. below the present at 10,000 B.P. The rise in sea-level is strong but is gradually levelling off after 6000 B.P.

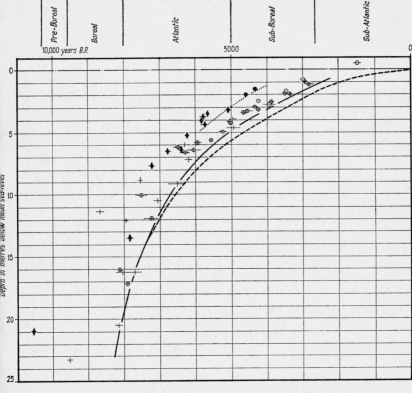

Fig. 1.6   *Curves of relative changes in sea-level in the Netherlands*
(after Jelgersma, 1961)

—⊕— Donken *(Rhine-Meuse estuary)*   —○— Swales *(coastal barriers)*
—+— Zuid *and* Noordholland   —●— Zeeland *(base Upper Peat)*
—◇— Friesland *and* Groningen   —— Curve I
—♦— Zeeland   ········ Curve II
                                              —— —— Curve III *(mean sea-level)*

Finally, the sea-level curve of Curray (1965) (slightly modified
after Shepard (1960) and Curray (1960)), thought to represent eustatic
fluctuation, is presented in Fig. 1.8. This curve gives an indication
of the movements of sea-level during the last 20,000 years. It indicates
that sea-level was 120 m. below its present level at 18,000 B.P. At
the beginning of the Holocene, the period under discussion in this
paper, sea-level reached a level of about 40 m. below the present.

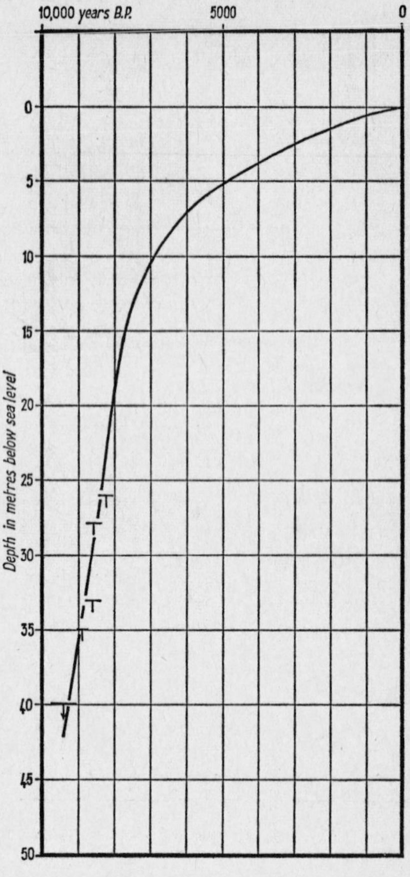

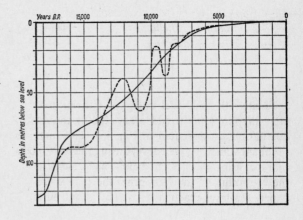

Fig. 1.7 Curve of relative changes in sea-level in the Netherlands and adjacent North Sea (after Jelgersma, 1961)

Fig. 1.8 Curve of sea-level changes (after Curray, 1965)

Thus the major restoration of ocean level after the last glaciation had already occurred when the Holocene period started. Before that time approximately two-thirds of the melting water of the glaciers had already been returned to the oceans.

The evidence of changes in level is derived from radiocarbon dates from shells of which the ecological position relative to sea-level is known. The samples are dredged from the shelf off the Texas coast, which is supposed to be tectonically stable.

This indicates that the constructed sea-level curve is supposed to reflect the eustatic movements of sea-level. The sea-level changes of Curray (1965), presented in Fig. 1.8, give two curves, the dotted curve estimated from minimal data and the solid curve the approximate mean of the compiled data. It must be mentioned that the fluctuations of sea-level shown by the dotted curve are for the greater part based on the characteristics of the sediments such as morphology and distribution on the Texas shelf (Curray, 1960). Regarding the reliability of the age of the regressions shown in his dotted curve, Curray (1961) states: 'The relict shoreline deposits can determine fairly precisely the heights of sea-level at the beginnings and ends of the brief regressions, but the dates are less certain.' It is thus possible that the ages of the regressions as indicated in the curve are not correct. This concerns especially the fall of sea-level that should have occurred about 9000 B.P. In our opinion this regression is questionable, because a growth of glaciers originated by a deterioration in climate is needed for such a phenomenon. A deterioration in climate during the period under consideration is unknown at the present time. The investigations of Curray have not given information about sea-level changes during the last 6,000 years as the studied shorelines observed by him originated before that time.

Summarising the results of the rising sea-level concept it must be mentioned that the data indicate a continuously rising sea-level throughout the Holocene, reaching its present position asymptotically only very recently. Tectonic movements could have influenced the derived results, as most of the studied areas may be subsiding slowly.

5. EUSTATIC SEA-LEVEL CHANGES

The preceding pages contain an outline of the studies on the sea-level changes of the last 10,000 years. The results obtained from the various studies are evaluated. The main problem in all investigated areas is the possibility of tectonic movements having taken place during the Holocene. Accordingly opinions are rather divergent, particularly concerning the last 5,000 years. From this part of the Holocene numerous former shorelines, submerged and emerged, have been investigated. These studies have given rise to three opinions on eustatic sea-level changes during the last 5,000 years: the steady, the rising and the oscillating sea-level concept.

Concerning the oscillating sea-level concept, which includes the high Post-glacial sea-level stages it must be stressed that this hypothesis is highly questionable. In the places observed, local warping must have had an important share in the changes in level. In other areas elevated shorelines once thought to be of Holocene age are by means of absolute age determination dated back to the Pleistocene.

The warmer climate during the Atlantic or Hypsithermal is not a proof that the sea-level at that period was higher than at present. Extensive studies on sediments of the last interglacial, the Eemian, by means of pollen analysis in France, England and the Netherlands, have given evidence that sea-level was still rising considerably after the climatic optimum of the Eemian (Zagwijn, 1961; West and Sparks, 1961).

If sea-level had been higher in recent times, about 3 m. as claimed by Fairbridge, coastal plains should have been inundated on a large scale. The data obtained from the intensively studied Holocene sections on the Gulf Coast, Florida, and the Netherlands lack evidence of such a high sea-level stage. There is of course the possibility that tectonic movements in the coastal plains exactly counteracted the high sea-level stages so that they do not show up in the presented curves. This hypothesis seems most unlikely, as it needs strongly increased tectonic subsidence during certain periods, occurring at the same time in the Gulf Coast region in Florida as well as in the Netherlands.

In the writer's opinion the evidences for high Post-glacial sea-level stages are questionable. Much more data are needed before the oscillating sea-level concept can be accepted.

At the present time the available data indicate a rising sea-level during the Post-glacial, slowing down after 6000 B.P. The present position was reached either 3,600 years ago or only very recently.

According to the curve of Scholl and Stuiver (1965), an eustatic rise of 1·6 m. should have occurred during the last 3,600 years. This seems to be an acceptable amount, as the steady sea-level concept is open to question. Also, in view of climatic fluctuations, it seems more likely that sea-level has been moving slightly instead of standing still during the last 3,600 years. Before final conclusions on this subject can be drawn, more investigations are needed.

6. HOLOCENE COASTAL SEQUENCES

It has been mentioned in the preceding pages that many data of changes in level are derived from submerged peat layers in coastal plains. Holocene sections in those areas consist of alternating peat, clay and sand layers of different thickness. The stratigraphy and lithology of the Holocene sequence on the coastal area of the Netherlands are summarised in Figs. 1.9 (*a*) and (*b*). The underlying Pleistocene and early Holocene on the section line consist of fluvial sediments deposited by the rivers Rhine and Meuse. In contrast to (*a*), the vertical scale of (*b*) represents the time of deposition of the sedimentary strata. It appears that thin peat layers in the Holocene may represent a considerable time interval. Because of compaction, the significance of thin peat layers in time may be underestimated as compared with sand and clay beds. The result is given in Fig. 1.9 based on field studies by members of the Geological Survey, supported by pollen analysis and numerous radiocarbon datings.

The presented data indicate that peat growth was interrupted at various times by marine transgressions. The whole Holocene sequence gives evidence of transgression and regression periods. In our opinion, however, important fluctuations of sea-level have not taken place in the Netherlands. Minor fluctuations may have occurred but cannot be recognised because of the sources of error, taken into account in constructing the sea-level curve (Jelgersma, 1961). Accordingly the words 'transgression' and 'regression' are not used here in the sense of a rise and fall of sea-level. By the word 'transgression' we mean only that the land was covered by the sea and by 'regression' that the sea retreated. At many places in the coastal area the times at which these so-called marine transgressions and regressions occurred can be established by radiocarbon datings of the peat directly below or directly on top of a marine deposit. Obviously these marine transgressions had only a limited extent, which depended on the nature of the shore, i.e. river mouths and inlets (Pons *et al.*, 1963).

The problem arises as to which phenomenon has caused this cyclicity. It could be caused either by eustatic fluctuation of sea-level or by a succession of wet and dry climatic conditions. In the latter

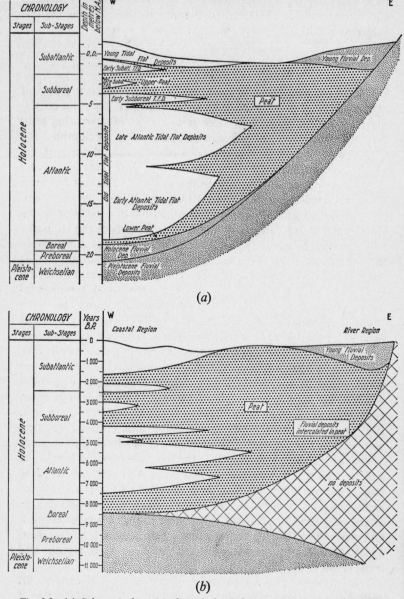

Fig. 1.9   (a) Schematical section through the Holocene in the western part of the Netherlands (after Jelgersma, 1961). (b) The same section plotted on a time scale.

case increased river activity during wet periods (more precipitation) together with an increase in storminess could be responsible for a so-called transgression period.

The relation of these regression and transgression phases to climatic fluctuations is a much discussed subject (Bakker, 1948; Bennema, 1954; Pons and Wiggers, 1959, 1960; Jelgersma, 1961). This hypothesis was first published by Gilette (1938), who accepted cycles of 567 years expressed in sedimentary stratification.

On the basis of the results of pollen analysis Iversen also accepted climatic fluctuations. A succession of wet and dry climatic conditions is represented by the so-called recurrence surfaces in European peat bogs. There are many recurrence surfaces (*Grenzhorizont*) of different ages, as indicated by pollen analysis and radiocarbon datings (Godwin and Willis, 1964).

During the last years geological investigations have been carried out in the coastal dunes (sub-Boreal and sub-Atlantic in age) of the Netherlands. Data obtained point to several phases of aeolian sand deposits separated from each other by peaty horizons. The idea is expressed that a dune region is a very unstable environment highly dependent on precipitation. Accordingly the peaty horizons are thought to represent wet periods. Radiocarbon datings of these horizons show them to be synchronous with the transgression periods in the back swamp. Therefore the aeolian phases should represent the regression periods. This investigation is not yet finished, and before final conclusions can be made we have to await more data.

There are, however, indications that the lithology of the Holocene sequence is influenced by a succession of wet and dry climatic conditions. It must be mentioned also that for the youngest part of the sequence, the sub-Atlantic period, a climatic succession may be recognised which looks very similar to that published by Lamb (1964).

The Holocene sequence of the Netherlands has a great similarity to sequences drived from investigations in north-west Germany (Brandt *et al.*, 1965) and the Fenlands of East Anglia (Willis, 1961).

An outline of the Holocene sequence in central coastal Louisiana, U.S.A., is published by Coleman and Smith (1964), and Fig. 1.10 represents a composite stratigraphic section as published by them. If we look upon the dated peat horizons, the resemblance to those dated in the Netherlands is striking. Coleman and Smith explain their section by a cyclic nature of sediment supply and/or rate of submergence. The sediment supply from the Mississippi river should depend on climatic fluctuations, and so, as in the Netherlands, the

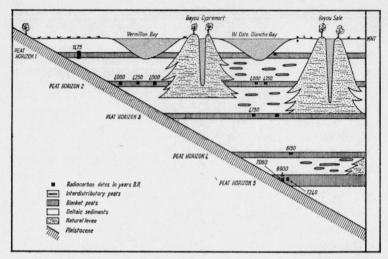

*Fig. 1.10   Composite stratigraphic section through the Holocene in central coastal
Louisiana (after Coleman and Smith, 1964)*

Holocene sequence in Louisiana should, to a large extent, be influenced by climatic fluctuations.

It is my opinion that an important contribution to the understanding of climatic fluctuation might be made by investigating and dating many coastal sequences in all parts of the world.

REFERENCES

BAKKER, J. P. (1948) 'Morfologisch onderzoek van Barradeel en zijn betekenis voor het inzicht in de subatlantische transgressie en het verspreidingsbeeld der terpen', *Kon. Ned. Akad. Wet.* (Amsterdam) II 121.

BENNEMA, J. (1954) 'Holocene movements of land and sea-level in the coastal area of the Netherlands', *Geol. Mijnb.*, n.s., XVI 254.

BLOOM, A. L., and STUIVER, M. (1963) 'Submergence of the Connecticut coast', *Science*, CXXXIX 332.

BRANDT, G., HAGEMAN, B. P., JELGERSMA, S., and SINDOWSKI, K. H. (1965) 'Die lithostratigraphische Unterteilung des marinen Holozäns an der Nordseeküste', *Geol. J.*, LXXXII 365.

BROECKER, W. S. (1965) 'Isotope geochemistry and the Pleistocene climatic record', in *The Quaternary of the United States*, INQUA VII Congress (Princeton, N.J.) p. 737.

—— and THURBER, D. L. (1965) 'Uranium-series dating of corals and oölites from Bahaman and Florida Key limestones', *Science*, CXLIX 58.

COLEMAN, J. M., and SMITH, W. G. (1964) 'Late recent rise of sea level', *Geol. Soc. Amer. Bull.*, LXXV 833.

CURRAY, J. R. (1960) 'Sediments and history of Holocene transgression, continental shelf, north-west Gulf of Mexico', *Amer. Assoc. Petr. Geol.* (1951–8) 221.

—— (1961) 'Late Quaternary sea-level: a discussion', *Geol. Soc. Amer. Bull.*, LXXII 1707.

—— (1965) 'Late Quaternary history, continental shelves of the United States', in *The Quaternary of the United States*, INQUA VII Congress (Princeton, N.J.) p. 723.

DALY, R. A. (1920) 'A recent world-wide sinking of ocean level', *Geol. Mag.*, LVII 246.

—— (1934) *The Changing World of the Ice Age* (Yale Univ. Press, New Haven).

EMERY, K. O., TRACEY, J. I., JR., and LADD, H. S. (1954) 'Geology of Bikini and nearby atolls. Pt. I: Geology', *U.S. Geol. Surv. Prof. Pap.* 260-A.

FAIRBRIDGE, R. W. (1961) 'Eustatic changes in sea-level', *Phys. Chem. Earth*, V 99.

FISK, H. N. (1951) 'Loess and Quaternary geology of the Lower Mississippi Valley', *J. Geol.*, LIX 333.

—— and MCFARLAN, E. (1955) 'Late Quaternary deposits of the Mississippi River', *Geol. Soc. Amer.*, Special Paper, LXII 279.

GILETTE, H. P. (1938) 'Coincidence of some climate and sea-level cycles', *Pan-Amer. Geologist*, LXX 279.

GILL, E. D. (1961) 'Change in the level of the sea relative to the land in Australia during the Quaternary era', *Z. Geomorphol. Suppl.*, III 73.

—— (1965) 'Radiocarbon dating of past sea-levels in south-east Australia', *Abstracts*, INQUA VII Congress (Boulder, Colorado, p. 167.

GODWIN, H., SUGGATE, R. P., and WILLIS, E. H. (1958) 'Radiocarbon dating of the eustatic rise in ocean level', *Nature*, CLXXXI 1518.

GOULD, H. R., and MCFARLAN, E. (1959) 'Geologic history of Chenier plain, south-western Louisiana', *Trans. Gulf Coast Assoc. Geol. Soc.*, IX.

GOURINARD, Y. (1958) I. 'Recherches sur la géologie du littoral Oranais', II. 'Épirogenése et nivellements', *Serv. Carte Géol. Alg.*, VI 1.

HAILS, J. R. (1965) 'The Late Pleistocene and Recent history of the Macleay Valley and adjacent coastal areas, mid-north coast of New South Wales', *Abstracts*, INQUA VII Congress (Boulder, Colorado) p. 187.

HOYT, J. H., WEIMER, R. J., and HENRY, V. J. JR. (1964) 'Late Pleistocene and recent sedimentation, central Georgia coast, U.S.A.', in *Developments in Sedimentology* (Elsevier, Amsterdam) I 170.

——, —— and —— (1965) 'Age of Pleistocene shoreline deposits, coastal Georgia', *Abstracts*, INQUA VII Congress (Boulder, Colorado) p. 228.

JELGERSMA, S. (1961) 'Holocene sea-level changes in the Netherlands', *Med. Geol. Sticht.*, series C–VI, 101 pp.

LAMB, H. H. (1964) 'Atmospheric circulations and climatic changes in Europe since 800 A.D.', *Report*, INQUA VI Congress (Warsaw, 1961) p. 291.

LE BLANC, R. J., and BERNARD, H. A. (1954) 'Résumé of late recent geological history of the Gulf Coast', *Geol. Mijnb.*, n.s. XVI 185.

MCFARLAN, E. (1961) 'Radiocarbon dating of Late Quaternary deposits, South Louisiana', *Geol. Soc. Amer. Bull.*, XXII 129.

MCKEE, E. D. (1959) 'Storm sediments on a Pacific atoll', *J. Sed. Petr.*, XXIX (3) 354.

MACNEIL, S. F. (1950) 'Pleistocene shorelines in Florida and Georgia', *U.S. Geol. Survey, Prof. Pap.* 221-F, p. 95.

PONS, L. J., JELGERSMA, S., WIGGERS, A. J., and JONG, J. D. DE (1963) 'Evolution of the Netherlands coastal area during the Holocene', *Verh. Kon. Ned. Geol.*, *Mijnbouwk. Gen., Geol. Serie*, XXI (2) 197.

PONS, L. J. and WIGGERS, A. J. (1959–60) 'De holocene wordingsgeschiedenis van Noordholland en het Zuiderzeegebied', I and II, *Tijd. Kon. Ned. Aardr. Gen.*, LXXVI 104; LXXVII 1.

REDFIELD, A. C., and RUBIN, M. (1962) 'The age of salt marsh peat and its relation to recent changes in sea-level at Barnstable, Massachusetts', *Proc. Nat. Acad. Sci.*, XLVIII 1728.

SCHOFIELD, J. C. (1960) 'Sea-level fluctuations during the past four thousand years', *Nature*, CLXXXV 836.

—— (1964) 'Post-glacial sea-levels and isostatic uplift', *N.Z. J. Geol. Geophys.*, VVI (2) 359.

SCHOLL, D. W. (1964) 'Recent sedimentary record in mangrove swamps and rise in sea-level over the south-western coast of Florida', Pt. 1, *Marine Geol.*, I 344.

—— (1965) ibid., Pt. 2, II 343.

—— and STUIVER, M. (1965) 'Recent submergence of south Florida: comparison with adjacent coasts and other eustatic data' (in press).

SHEPARD, F. P. (1960) 'Rise of sea-level along north-west Gulf of Mexico: recent sediments', *Amer. Assoc. Petr. Geol.* (1951–8) 338.

—— (1963) 'Thirty-five thousand years of sea-level', in *Essays in Marine Geology* (Univ. S. Calif. Press, Los Angeles) 201 pp.

—— (1964) 'Sea-level changes in the past 6,000 years; possible archaeological significance', *Science*, CXLIII 574.

—— and CURRAY, J. R. (1965) 'C¹⁴ determinations of sea-level changes in stable areas', *Abstracts*, Internat. Assoc. Quatern. Research, VII International Congress, p. 424.

—— and SUESS, H. E. (1956) 'Rate of post-glacial rise of sea-level', *Science*, CXXIII 1082.

STRAATEN, L. M. J. U. VAN (1954) 'Radiocarbon datings and changes of sea-level at Velzen (Netherlands)', *Geol. Mijnb.*, n.s., XVI 247.

TESCH, P. (1947) 'De niveauveranderingen en de oudheidkundige verschijnselen', in *Een kwart eeuw oudheidkundig bodemonderzoek*, Gedenkboek A. E. van Giffen, p. 43.

THURBER, D. L., BROECKER, W. S., POTRASZ, H. A., and BLANCHARD, R. L. (1965) 'Uranium series ages of coral from Pacific atolls', *Science* (in press).

WARD, W. T. (1965) 'Eustatic and climatic history of the Adelaide area, south Australia', *J. Geol.*, LXXIII 592.

WEST, R. G., and SPARKS, B. W. (1961) 'Coastal interglacial deposits of the English Channel', *Phil. Trans. Roy. Soc. B* (London) CCXLIII 95.

WILLIS, E. H. (1961) 'Marine transgression sequences in the English Fenlands', *Ann. N.Y. Acad. Sci.*, XCV (1) 368.

ZAGWIJN, W. H. (1961) 'Vegetation, climate and radiocarbon datings in the Late Pleistocene of the Netherlands', Pt. I, *Med. Geol. Sticht.*, n.s., XIV 15.

2 The Main Post-glacial Raised Shoreline of Western Scotland from the Firth of Lorne to Loch Broom

S. B. McCANN

THE various inter-glacial, Late-glacial and Post-glacial raised beaches of western Scotland provide evidence of marine action at different levels above present sea-level up to a maximum height of 135 ft recorded in Colonsay, and have been regarded as representing a number of distinct raised shorelines. In this sequence the features which have been grouped together as the Post-glacial 25-ft beach are the most striking, and their continuous presence in many parts of the area is the most remarkable feature of a coastline with many features of special geomorphological interest. An assessment of their height variation throughout the area is of interest, therefore, not only in relation to the particular problems of the isostatic readjustment of formerly glaciated regions and the associated eustatic changes in sea-level, but also as a contribution to the understanding of the development of the coastal landforms of the area in question.

There is a considerable volume of literature concerning the raised beaches and related deposits of Scotland, and a full review even of that section of it dealing with the Post-glacial beaches is not appropriate in this paper. It is necessary, however, to place the present work in its proper context by attempting to establish how there came to be formulated the widely accepted view that it is possible to recognise a single major raised shoreline of Post-glacial age throughout a wide area embracing not only most of Scotland but also northeast Ireland and northern England. In this respect the writings of W. B. Wright are most important, particularly his influential volume, *The Quaternary Ice Age*, published in 1914, which contains the first clear statement of the generalisation outlined above, together with a map, previously published in 1911, of the extent or zero isobase of this shoreline. Although one can cite earlier descriptions of the 25-ft raised beach of Scotland, for instance that in A. Geikie's *Scenery of*

Scotland (1865), and can clearly recognise the debt owed to the work of Jamieson (1865), Wright's lucid generalisations provide the first overall synthesis of information relating to the extent and height variation of this shoreline.

As an officer of the Geological Survey, with wide field experience and a special interest in raised beach problems, Wright (1904, 1911) was in a good position both to summarise the state of knowledge and to influence the thinking of his colleagues who were working in western Scotland, preparing the one-inch sheets of the Geological Map. Fifteen of these one-inch maps and their accompanying *Memoirs*, covering the greater part of western Scotland from Arran and south Bute to Ross-shire, including the islands of Islay, Jura, Colonsay and Mull, were published in the period 1903–25. Wright himself was part author of five of them (covering Knapdale, Colonsay and Oronsay, Mull, Staffa and Iona, and Ben Nevis and Glencoe), the most important being the 1924 memoir on the Tertiary and post-Tertiary geology of Mull, which contains the fullest account of raised beaches and related phenomena of any of the Survey publications. All the memoirs, however, contain some brief descriptions of the raised beaches, and their extent is indicated on all the one-inch maps, thus providing the most complete record of these features for the greater part of western Scotland, a record which leaves little doubt in the reader's mind that there is a major Post-glacial raised shoreline, the 25-ft raised beach, which can be recognised throughout the area, at progressively lower elevations away from Loch Linnhe. Thus these descriptions of field evidence from western Scotland accord with, and give validity to, Wright's generalisations concerning the Post-glacial raised beach, which are repeated in a paper of 1928, and in the second edition, of 1937, of *The Quaternary Ice Age*.

Confining one's attention to western Scotland, and more particularly to that part of the west coast which is, the subject of this paper two later contributions by McCallien (1937) and Donner (1959) are important, the first for its suggestion that not all the features which have been considered together as the Post-glacial raised beach were formed during the short period of the Post-glacial submergence, and the second for its view that the shoreline is horizontal and not tilted. These are major departures from Wright's view and will be considered below, under two separate headings, together with information from outside western Scotland which is relevant to the discussion. The problem of what may properly be considered to represent raised shoreline features of Post-glacial age, raised by McCallien, is the more important and will be considered first, together with brief descriptions of these features in the area under consideration.

RAISED SHORELINE FEATURES OF POST-GLACIAL AGE

The raised shore features which have been grouped together as the 25-ft beach in the maps and memoirs of the Geological Survey include almost every kind of form – shore platforms and cliffs, with caves, cut in rock; shore terraces and cliffs cut in older drift deposits; storm-beach ridges; shingle and sand beaches and deltas – but it is the erosional forms in rock which are problematical in terms of the length of time necessary to produce such well-defined features. Wright called the 25-ft sea the cliff maker par excellence, and recognised that some of the features which have been associated with it represent an amount of erosion exceeding that on the present shore. He was therefore of the opinion that 'it is fairly certain that its formation occupied a longer period than has elapsed since it was raised to its present position'. McCallien, considering the chronology of the Post-glacial period, doubted that there had been sufficient time for the erosion of so much solid rock. He suggested instead that the rock platforms and cliffs associated with the 25-ft beach are much older, of Pre-glacial or inter-glacial age, and that marine action during the Post-glacial submergence had inherited at something like its maximum level an already existing series of coastal features cut in rock.

In the absence of any subsequent positive contribution from western Scotland this suggestion has remained speculative, but has been given greater validity by Stephens's (1957) descriptions of the relationships of the Post-glacial raised beach and a much older rock platform from the east coast of Ireland. The low rock platform, which occurs at intervals along the coast as far north as Belfast, though without any discernible sign of tilting, is overlain in places by two separate glacial drifts, and it has been inferred that its formation may be placed at least as far back as the Mindel–Riss Inter-glacial. The Post-glacial shoreline, which declines in height southwards at the rate of one-tenth of a foot per mile, from 34–37 ft above Irish O.D. in north Antrim to 22–24 ft in Dublin Bay, has a well-marked cliff-line cut in glacial drifts, which in places have been stripped back so that Post-glacial raised beach deposits rest directly upon the much older rock platform. There is nowhere any definite indication that the Post-glacial sea cut any significant platforms in rock (Stephens and Synge, 1965). It seems unlikely, therefore, that this can have been the case in western Scotland. In view of the Irish evidence it seems highly probable that the well-developed raised shore platforms cut in rock are much older features which have been exhumed and trimmed by the Post-glacial sea, though positive proof does not appear to be forthcoming.

Within the area described in this paper, and in the islands of Islay and Jura to the south-west, the author has investigated many miles of coastline with a view to demonstrating this point, without being able to discover a coastal section where the rock platform can be seen to to be overlain by undoubted glacial drift. None the less, one records from many parts of the area repeated pieces of circumstantial evidence (such as the appearance of ice-moulding of the rock platform, the existence of striae-like markings at the seaward margin of the platform, and the presence of thin veneers of what appears to be glacial drift overlying the platforms) which, taken together, are convincing.

This evidence relates only to the rock platforms, for the cliffs at the rear are always freshlooking, with the appearance of very recent erosion. They are usually vertical or near-vertical, with features such as caves and undercut notches indistinguishable in form from those of the modern shore. There can be no doubt that these cliffs owe much of their present form to marine erosion during the Post-glacial raised-beach period. Erosion during this period was sufficiently prolonged not only to remove all traces of glacial drift from the rock cliffs, which originally backed the old rock platforms, but also to carry out the significant amount of erosion which was responsible for the almost contemporary nature of the raised cliff-line in rock. This erosion has caused little alteration of the platforms, however, and on occasion has not quite removed all the glacial drift, which may remain as a veneer or in isolated pockets.

The rock platforms, which are considered above to be a much older feature than other shore forms associated with the Post-glacial raised beach, are not present throughout the whole of western Scotland, nor does the significant Post-glacial cliffing of solid rock occur throughout the area. The distribution of the two phenomena are, however, not the same. Well-developed rock platforms do not occur north of a line joining Fort William, Tobermory and Colonsay, but significant cliffing of solid rock during the Post-glacial raised-beach

period is found as far north as Kyle of Lochalsh on the mainland
and Isle Ornsay in Skye. E. B. Bailey (Bailey *et al.*, 1924) has
commented on the contrasted development of the Post-glacial beach
between the north-west and south-west Highlands, but did not
distinguish between the two types of feature, considering both
platform formation and rock cliffing to have been achieved during
the Post-glacial raised-beach period. Consequently his limit of
pronounced marine erosion during this period is substantially the
same as the northern limit of wide, well-developed rock platforms.
The explanation offered by Bailey of the difference between the two
areas, as regards platform development, is that the south-west
enjoyed a more prolonged period of constant sea-level during the
Post-glacial submergence than did the north-west. In the present
author's view this explanation would only account for the progressive
decrease, in a northerly direction, of the amount of cliff cutting
associated with the Post-glacial raised beach: the presence of wide,
well-developed platforms in the south-west must be attributed to
the existence in that area of a much older feature the level of which
coincided with that of the Post-glacial raised beach.

Within the area considered in this paper, raised rock platforms
are well developed in the Firth of Lorne – Loch Linnhe area,
especially along the mainland coastline from Oban to Kentallen,
around the Islands of Kerrera, Lismore, and Shuna, and in eastern
and south-eastern Mull. Here the platforms, normally 20–30 yds
wide, though occasionally much wider, have a gently sloping surface,
which gives way seawards, by way of a rounded margin, to the
vertical or near-vertical cliff faces which reach down to present high-
water mark. Although contemporary marine action is involved in
cutting these low cliffs, which may be 10–15 ft high, there is no
significant development of modern shore platforms. The raised
platform, especially in the outer part (which is bare of vegetation)
and in the present spray zone, may show the appearance of ice-
moulding and striations. These features are best seen in northern
Lismore and eastern Mull, especially along the mile or so of coast
north and south of the entrance to Loch Spelve. In this connection
it is interesting to note the following remark, printed on the one-inch
map of eastern Mull, sheet 44, opposite Portfield, south of Loch
Spelve: 'Little cliff below raised beach platform often retains ice
markings north of this but is everywhere wave-worn to the south.'
The thin veneers of drift material, which may represent glacial
material resting on the platform, are less convincing than the

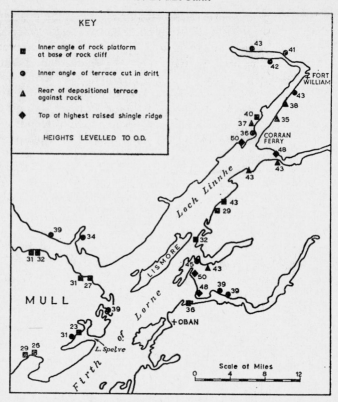

KEY

■ Inner angle of rock platform
    at base of rock cliff

● Inner angle of terrace cut in drift

▲ Rear of depositional terrace
    against rock

◆ Top of highest raised shingle ridge

HEIGHTS LEVELLED TO O.D.

*Fig. 2.1   Post-glacial raised shoreline features along the Firth of Lorne and
Loch Linnhe*

indications of the erosive effect of ice; they are perhaps best examined
along the west coast of Loch Linnhe north of Corran Ferry, though
similar deposits occur in western Lismore. It is possible also that the
fluvio-glacial gravels of that part of the Corran fan on the east side
of Loch Linnhe (McCann, 1961), are banked against this platform,
but the relationships are not clear.

It is these platforms in the areas so far described which the author
regards as having been exhumed by Post-glacial marine action, which
retrimmed the cliff at the rear but had little effect on the platform
itself. Their height at a number of localities is shown in Fig. 2.1,
together with heights of certain wholly Post-glacial shoreline features,
which are grouped into three types – erosional terraces cut in drift,
depositional terraces banked against rock, and storm-beach ridges.

Two points of note are illustrated by this map. First, there is a discernible decrease in the height of all these features in a south-westerly direction. Secondly, in any one part of the area the height relationships between the four sets of features appear to accord with those one might expect to find between the corresponding features being formed contemporaneously on the modern shoreline. Highest, at 48–50 ft above O.D., are the former storm-beach ridges which developed in the areas of abundant supply of beach material at the sites of the fluvio-glacial gravel fans marking the limit of the Highland or Loch Lomond readvance (McCann, 1961), which by chance are located in areas of maximum exposure to wave action. Some 8–10 ft below these shingle ridges are the depositional beach terraces and beach terraces eroded in drift material: the rock platforms are usually slightly below this again. Thus, whatever evidence there may be of a much greater age for the raised rock platforms of the area, there is no indication in terms of height that they differ significantly from the other shore features with which they have been grouped as the main Post-glacial raised shoreline.

Away from the Firth of Lorne and Loch Linnhe, in the north and west, raised rock platforms are poorly developed; and, where they do occur, with associated rock cliffs, the amount of erosion is always readily attributable to Post-glacial marine action. This wholly Post-glacial erosion of solid rock is best seen along the coast for some 10–12 miles south of Mallaig, though similar evidence of cliff formation during the Post-glacial submergence occurs at numerous localities as far north as Kyle of Lochalsh. Typically, vertical or near-vertical cliff faces, exhibiting caves, undercut notches, geos and other features indicative of relatively recent erosion, are separated from the sea by a low broken zone, often only 4 or 5 yds wide, which represents a very rudimentary shore platform. The slope of this rudimentary platform, which often has numerous stacks and a considerable cover of raised beach gravel, is continuous with that of the present shore, there being no separating cliff between the platform and high-water mark.

There is no evidence along the mainland coast north of Ardnamurchan which would indicate the presence, above high-water mark of an older rock platform such as was inherited by the Post-glacial submergence in the Firth of Lorne area; consequently the Post-glacial shoreline lacks the continuity which characterises it in the south-west. It is represented by low rock cliffs and rudimentary shore platforms wherever rock type and exposure to wave attack made their development possible in the short period of time available; and by low terraces cut into older drift deposits, wherever this material is present along the coast. There are also occasional depositional forms, such as low isolated rounds of raised beach gravel and raised deltas, but these are unsuitable for height determination. This pattern of development is true also of western Mull and of the island of Skye, though at certain localities in these two islands there are complications introduced by the reappearance of rock platforms at a very low level which may be considerably older than the Post-glacial submergence.

The mainland coast north from Kyle of Lochalsh to Loch Broom exhibits no sign of the erosion of solid rock during the Post-glacial submergence, and the Post-glacial shoreline is represented by terraces cut in older drift, which may be boulder clay as in the area north and south of Loch Gairloch, fluvio-glacial gravels as at the entrance to Loch Carron, or raised deltas associated with Lateglacial raised beaches as at Ullapool in Loch Broom. In all cases the terraces are narrow, and continuous in profile with the present shore.

In concluding this discussion of the nature of the raised shore features of Post-glacial age in western Scotland, which used as a starting point McCallien's suggestion that the Post-glacial 25-ft raised beach may have inherited the rock platform of an earlier shoreline, the author's views may now be summarised:

1. Along the west coast of Scotland between Oban and Ullapool, including the islands of Mull and Skye, it is possible to recognise a major Post-glacial shoreline – the 25-ft raised beach of the literature.
2. Though it is possible to discern, below the level of the main shoreline in certain localities, features which were formed during the fall in sea-level from the Post-glacial maximum, these do not appear to constitute a separate shoreline which can be recognised throughout.
3. The amount of erosion of solid rock associated with the formation of the main raised shoreline was nowhere sufficient to produce well-developed shore platforms, though a certain amount of cliffing did take place with associated very rudimentary platform development.
4. This relatively small amount of erosion of solid rock during the Post-glacial submergence decreases northwards and

westwards away from the Firth of Lorne–Loch Linnhe area, ceasing altogether on the mainland coast north of Kyle of Lochalsh.

5. The conspicuous raised shore platforms in rock occurring in the south-east of the area – in eastern Mull and along the Firth of Lorne and Loch Linnhe – at a similar height to that of truly Post-glacial shoreline forms, must be regarded as much earlier features, which were inherited by the Post-glacial sea at something like its maximum level, with consequent renewed cliffing at the rear but little alteration of the existing platforms.

CRITICISM OF J. J. DONNER'S VIEWS

The shoreline diagrams presented by Donner (1959), and the conclusions drawn from them concerning the classification and height variation of both the Late- and Post-glacial shorelines of Highland Scotland, are based on height measurements at only thirty sites, apparently haphazardly selected, around the coast from Loch Long to the Moray Firth, only nine of which fall within the area considered in this paper. The data are clearly inadequate and the interpretation reveals an insufficient appreciation of the problems involved. It is not possible to trace the height variation of any of the raised shorelines of Scotland simply by taking spot checks at widely isolated localities. This fact is perhaps best illustrated by reference to Donner's interpretation of the single measurement he made in the Loch Linnhe area, of a gravel terrace at Ballachulish, which he records as 15 m. above the barnacle line on the present shore, and which the present author records as 43·3 ft above O.D. (see Table 2.1, site 10). This terrace Donner considers to be part of a 50-ft Late-glacial shoreline, but even a brief reconnaissance of the surrounding area would have indicated that this view is untenable; the terrace is at a similar height to other raised shoreline features along Loch Linnhe and the Firth of Lorne, which clearly represent the main Post-glacial shoreline (see Fig. 2.1).

The incorrect interpretation placed by Donner on the age and field relationships of the Ballachulish terrace is sufficient to destroy his case that the main Post-glacial shoreline is horizontal throughout western Scotland, and further detailed comment on his interpretations of the height data he obtained from the eight other sites between Loch Linnhe and Loch Broom is unnecessary. As a useful contribution to understanding the general problems associated with the

TABLE 2.1
Details of height measurement sites

Site No.	Location	National Grid reference	Modified Grid reference		Height
			x	y	z
1	Loch Eil	NN 008784	100	078	42·6
2	Loch Eil	NN 043772	104	077	42·4
3	Loch Eil	NN 066778	106	077	41·1
4	Loch Linnhe	NN 083717	108	071	43·1
5	Loch Linnhe	NN 065691	106	069	37·7
6	Loch Linnhe	NN 022665	102	066	39·9
7	Loch Linnhe	NN 044663	104	066	34·6
8	Loch Linnhe	NN 011653	101	065	37·4
9	Loch Linnhe	NN 014638	101	063	36·4
10	Loch Leven	NN 055595	105	059	43·3
11	Loch Linnhe	NN 003579	100	057	43·3
12	Loch Linnhe	NM 953521	095	052	43·0
13	Loch Linnhe	NM 942512	094	051	28·9
14	Loch Linnhe	NM 902453	090	045	31·7
15	Loch Creran	NM 908415	090	041	44·5
16	Loch Creran	NM 928409	092	040	43·3
17	Loch Etive	NM 943359	094	035	39·4
18	Loch Etive	NM 964350	096	035	38·8
19	Dunstaffnage Bay	NM 896344	089	034	35·6
20	Mull, Loch Don	NM 742326	074	032	38·9
21	Mull, Sound of Mull	NM 710388	071	038	27·4
22	Mull, Sound of Mull	NM 692389	069	038	31·3
23	Mull, Sound of Mull	NM 613433	061	043	33·2
24	Mull, Sound of Mull	NM 605433	060	043	31·3
25	Mull, Calgary Bay	NM 359496	035	049	27·0
26	Mull, Loch Tuath	NM 394454	039	045	20·1
27	Mull, Loch Tuath	NM 453409	045	040	31·5
28	Mull, Loch na Keal	NM 524408	052	040	21·5
29	Mull, Loch na Keal	NM 529385	052	038	25·3
30	Mull, Loch na Keal	NM 509371	050	037	29·3
31	Mull, Loch Scridain	NM 491287	049	028	22·0
32	Mull, Loch Scridain	NM 506291	050	029	22·0
33	Mull, Loch Scridain	NM 519298	051	029	22·0
34	Mull, Loch Scridain	NM 529278	052	027	21·5
35	Mull, Loch Scridain	NM 471245	047	024	18·1

TABLE 2.1 (*continued*)

Site No.	Location	National Grid reference	Modified Grid reference		Height
			x	y	z
36	Mull, Loch Scridain	NM 411234	041	023	19·9
37	Mull, Loch Buie	NM 594243	059	024	28·6
38	Mull, Loch Buie	NM 602247	060	024	25·8
39	Mull, Loch Spelve	NM 682284	068	028	31·5
40	Morvern, Sound of Mull	NM 641457	064	045	38·8
41	Morvern, Loch Aline	NM 693457	069	045	33·9
42	Ardnamurchan, Kilchoan	NM 475633	047	063	25·9
43	Arisaig	NM 614836	061	083	19·4
44	Arisaig	NM 622849	062	084	17·8
45	Arisaig	NM 640853	064	085	18·7
46	Morar, Portnaluchaig	NM 652894	065	089	25·4
47	Morar, Traigh House	NM 658908	065	090	25·8
48	Morar, Glasnacardoch Bay	NM 672954	067	095	28·7
49	Skye, Isle Ornsay	NG 704115	070	111	25·7
50	Skye, Kyleakin	NG 725263	072	126	30·4
51	Skye, Rudha na Sgrianadin	NG 628260	062	126	25·6
52	Skye, Balmeanoch Bay	NG 532349	053	134	27·9
53	Skye, Staffin	NG 479682	047	168	21·9
54	Skye, Uig	NG 393633	039	163	19·9
55	Skye, Dunvegan	NG 246475	024	147	15·1
56	Skye, Loch Caroy	NG 303438	030	143	18·7
57	Skye, Bracadale	NG 339374	033	137	21·6
58	Skye, Glenbrittle	NH 398199	039	119	23·1
59	Loch Alsh, Balmacara	NG 846268	081	126	18·8
60	Loch Carron, Portchullin	NG 844346	084	134	26·9
61	Applecross	NG 711446	071	144	22·8
62	Applecross	NG 708457	070	135	26·8
63	Opinan	NG 742726	074	172	22·3
64	Loch Gairloch	NG 791773	079	177	24·7
65	Melvaig	NG 738861	073	186	21·2
66	Poolewe	NG 853810	085	181	18·9
67	Aultbea	NG 873887	087	188	15·8
68	Gruinard Bay	NG 945900	094	190	14·8
69	Gruinard Bay	NG 961922	096	192	15·2
70	Loch Broom	NH 124938	112	193	13·1

vertical displacement of raised shorelines in Scotland, Donner's paper fails because of the inadequacy of the data: as a contribution to understanding, in particular, the problems associated with the main Post-glacial shoreline the paper fails because Donner ignored, apart from the one isolated measurement at Ballachulish, the whole of the Loch Linnhe–Firth of Lorne area. This omission is difficult to understand, for this is the area where the shoreline attains its finest development, and where it has generally been accepted that it reaches its greatest elevation in western Scotland.

THE TILTING OF THE MAIN POST-GLACIAL SHORELINE

Measurements of the height of the main Post-glacial shoreline at seventy sites throughout western Scotland, between the Firth of Lorne and Loch Broom, are given in Table 2.1 (column 6) and their location shown on Fig. 2.2. The feature measured in each case is the inner angle between the raised cliff-line and the rear part of raised terraces and shore platforms of erosional origin; erosional features in both rock and older drift are included. The figures are given in feet above O.D. Newlyn and were determined by precise levelling.

Each figure is an actual field measurement at one selected locality which is considered to be representative of the height of the shoreline within the kilometre square of the grid reference (Table 2.1, column 3): it is not a mean value of a series of measurements. The normal procedure in levelling was:

1. to select a section of coast where the Post-glacial shoreline is well developed as an erosional feature and within convenient surveying distance of an O.S. bench mark;
2. to measure the height of the inner angle of the raised terrace along the coastline for a distance up to 200 yds depending on local circumstances of development (in many areas little variation in height is recorded along the coast; in others there are considerable local variations in height);
3. to select one specific locality, on the basis of the evidence from (2), which is representative in terms of height and development of the Post-glacial shoreline along the section of coast under consideration. The height recorded at this specific locality is the one which is listed in the table.

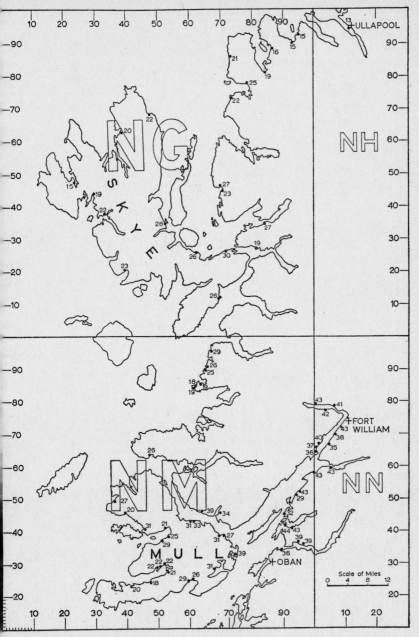

Fig. 2.2 The coastline of western Scotland from the Firth of Lorne to Loch Broom, showing the location of the measurement sites and National Grid co-ordinates. Height figures are given to the nearest foot above O.D.

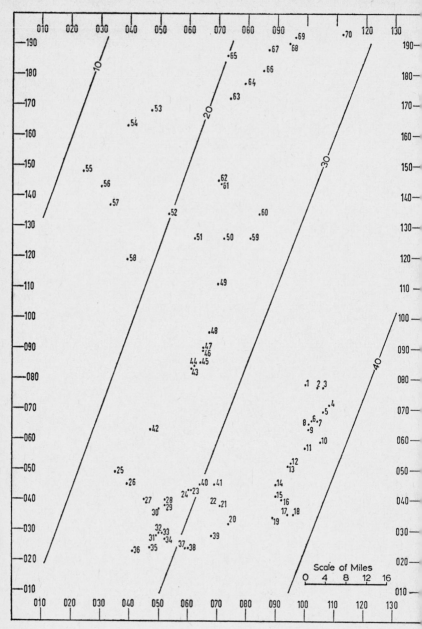

Fig. 2.3   Measurement sites (numbered 1–70) with modified grid co-ordinates and the position of the computed plane indicated

The National Grid reference is given in the normal manner correct to the nearest tenth of a kilometre, in order that each measurement site may be precisely located in the field. However, for the purposes of the analysis carried out below, these grid references are unsatisfactory, as the National Grid system involves a letter code for the major 100-km. squares and separate numbering of the eastings and northings within each major square (Fig. 2.2); accordingly, that part of the National Grid system involving 100-km. squares NM, NN, NG and NH was converted to a simple consecutively numbered grid system, without a letter code, with its origin at the south-west corner of square NM (Fig. 2.3). The converted x (eastings) and y (northings) co-ordinates of the measurement sites are given in columns 4 and 5 of the table, correct to the nearest kilometre.

Inspection of the height data contained in Table 2.1 and shown in location on Fig. 2.2 indicates that the main Post-glacial shoreline, as defined by the author, declines in altitude in all westerly directions away from Loch Linnhe. Along Loch Linnhe raised shore features attributed to this shoreline are frequently recorded at heights above 40 ft O.D.; from this high level they fall to below 20 ft in western Mull to the south-west, and to a similar height in western Skye to the north-west.

Using the new grid co-ordinates (x and y) and the height information obtained at each locality (z) it is possible to characterise, in a general manner, the tilting of the shoreline throughout the area by fitting a simple plane to the data. The plane is of the form $ax + by + z + c = 0$, and with the values of a, b and c calculated it becomes

$$-0 \cdot 2250 \times + 0 \cdot 0882y + z - 19 \cdot 5819 = 0.$$

This equation represents a plane inclining in the direction N.68° 35′W., at the rate of 0·396 ft per mile, which information is illustrated on Fig. 2.3 by drawing in the strike lines of the plane at 40 ft, 30 ft, 20 ft and 10 ft. (The north point referred to is grid, and not magnetic north.) To check how far the calculated plane adapts itself to the given points, it is necessary to correlate the sum of the squared deviations between the co-ordinates of the given points and the corresponding co-ordinates according to the equation of the plane. The square mean value of the deviation, d, is ±5·464 ft. Having derived the direction of maximum tilt of the shoreline, as represented by a single plane, the information may also be illustrated by a shoreline diagram, with the heights projected on to a base line drawn in this direction (Fig. 2.4 (a)).

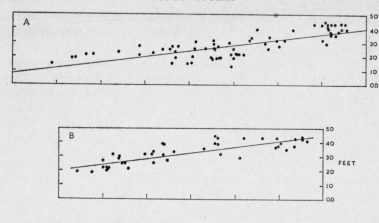

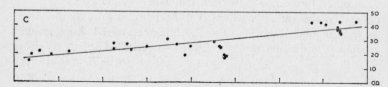

Fig. 2.4   Shoreline diagrams of the main Post-glacial shoreline between the Firth of Lorne and Loch Broom

A. Using height data from all 70 sites, projected on to a base line drawn in the direction of maximum tilt of the plane illustrated in Fig. 2.3.
B. Using height data from sites 1–42, projected on to a base line joining sites 4 and 36 (regression equation $y = 39 \cdot 0153 - 0 \cdot 3083x$; $r = 0 \cdot 8339$).
C. Using height data from sites 1–10 and 43–62, projected on to a base line joining sites 4 and 55 (regression equation $y = 43 \cdot 5817 - 0 \cdot 4242x$; $r = 0 \cdot 8484$).
In all diagrams the base line is divided into units of 10 miles.

This method of characterising and illustrating the height information relating to the Post-glacial shoreline in the area between the
Firth of Lorne and Loch Broom is adequate enough to demonstrate
the tilting of the shoreline, which is apparent from a simple inspection of the data, and provides values for the generalised direction
and amount of tilt. It remains, however, very much of a generalisation, for the differential amount of tilting of the shoreline in a
south-westerly direction, as compared with a north-westerly direction
away from Loch Linnhe, becomes obscured in the overall view. That
there is such differential tilting is illustrated by the other two shoreline diagrams in Fig. 2.4. The first of these (Fig. 2.4 (*b*)) was constructed using the height data from sites 1–42, that is those sites to
the south-west of Loch Linnhe along the Firth of Lorne and in

Mull, projected on to a line joining sites 4 and 36; the regression line representing the shoreline slopes away from Loch Linnhe at the rate of 0·424 ft per mile. The other diagram (Fig. 2.4 (c)) was constructed using the height data from sites 1–10 and 43–62, that is those sites to the north-west of Loch Linnhe along the mainland coast as far north as Applecross and in Skye, projected on to a line joining sites 4 and 55; the regression line representing the shoreline in this case slopes away from Loch Linnhe at the rate of 0·306 ft per mile.

CONCLUSIONS

The author's observations concerning the general nature of Post-glacial raised shoreline features in western Scotland between the Firth of Lorne and Loch Broom have been summarised, and the view put forward that it is possible to recognise throughout the area a single raised shoreline, the 25-ft raised beach of the literature. In concluding one may add the fact that the main Post-glacial shoreline declines in height in all westerly directions away from Loch Linnhe. If all the height measurements are considered together, the overall tilting of the shoreline is in a WNW. direction at the rate of 0·396 ft per mile, but there is good evidence of differential tilting of the shoreline, especially as between a south-westerly and a north-westerly direction away from Loch Linnhe.

REFERENCES

BAILEY, E. B. *et al.* (1924) 'The Tertiary and Post-Tertiary geology of Mull Loch Aline and Oban', *Mem. Geol. Survey of Great Britain.*

DONNER, J. J. (1959) 'The Late- and Post-glacial raised beaches in Scotland', *Ann. Acad. Sci. Penn.*, A 111, *Geol.-Geogr.*, LIII 1–25.

JAMIESON, T. F. (1865) 'On the history of the last geological changes in Scotland', *Q.J. Geol. Soc.*, XXI 161–203.

McCALLIEN, W. J. (1937) 'Late-glacial and early post-glacial Scotland', *Proc. Soc. Antiquaries Scotland*, LXXI 174–206.

McCANN, S. B. (1961) 'Some supposed raised beach deposits in Corran, Loch Linnhe and Loch Etive', *Geol. Mag.*, LXXXIX 131–42.

STEPHENS, N. (1957) 'Some observations on the "interglacial" platform and the early Post-glacial raised beach on the south coast of Ireland', *Proc. R. Ir. Acad.*, LVIII-B, 129–49.

—— and SYNGE, F. M. (1965) 'Late Pleistocene shorelines and drift limits in north Donegal', *Proc. R. Ir. Acad.*, LXIV-B, 131–53.

WRIGHT, W. B. (1911) 'On a pre-glacial shoreline in the western isles of Scotland', *Geol. Mag.*, XLVIII 97–109.

—— and MUFF, H. B. (1904) 'The Pre-glacial raised beach of the south coast of Ireland', *Sci. Proc. R. Dublin Soc.*, X 250–324.

ICD C

Trend-Surface Mapping of Raised Shorelines

S. B. McCANN AND R. J. CHORLEY

THERE have been several attempts in recent months to illustrate and define the degree of tilting or deformation of the various raised shorelines in Scotland. Considerable use has been made of shoreline distance diagrams or graphs, whereby heights of the raised beaches in a particular area are projected on to a base line, usually drawn in the supposed direction of maximum tilt of the series of shorelines as a whole. Each separate raised beach height is shown as a point on the diagram, the position of which is defined by altitude (usually given above O.D.) and projected distance along the base line: the shorelines are depicted by lines or zones on the diagram, which in some cases are calculated best-fit regression lines to the series of points representing a given shoreline; in other cases they have been fitted by eye.

In the sense that such diagrams are two-dimensional representations of height data which have areal distribution, they provide a partial picture and allow only limited analysis. A three-dimensional view may be obtained by fitting a simple plane to the data for each shoreline as has been done in the case of the main Post-glacial shoreline in part of western Scotland (McCann, 1966). More sophisticated analysis is possible, however, if higher order surfaces are considered. Techniques and applications of trend-surface mapping have been considered by Chorley and Haggett (1965), and the surface fitting programme described by Whitten (1963) offers a convenient tool in the present context. As will be illustrated later, trend-surface mapping provides a most useful method of presenting and analysing height data relating to raised shorelines. Regional trends may be precisely defined and an isobase map readily constructed. Local anomalies are revealed which raise further questions concerning the nature and causes of the deformation of the shorelines, and which in some cases may raise doubts as to whether the original assumption, that one is dealing with the same shoreline throughout, is correct.

Best-fit trend surfaces of linear, quadratic and cubic form have been fitted to the data recorded by McCann (1966) for the main Post-glacial shoreline in western Scotland, between the Firth of Lorne and Loch Broom, by the electronic computer programme described by Whitten (1963), using an I.B.M. 7090 computer. The

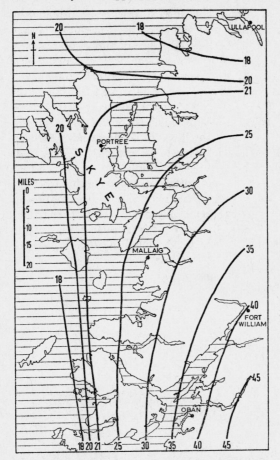

Fig. 2.5 Best-fit quadratic surface fitted to height data for the main Post-glacial shoreline in western Scotland, between the Firth of Lorne and Loch Broom. Figures are given in feet above O.D.

best-fit linear surface which had been fitted to the data previously, without the use of a computer, explains 60·27 per cent of the observed variation in height; the best-fit quadratic surface raises the level of explanation to 76·97 per cent, and the best-fit cubic surface to 78·23 per cent. In view of the small increase in explanation provided by the cubic as compared with the quadratic surface, the latter is considered most appropriate for further analysis.

This best-fit quadratic surface (Fig. 2.5) indicates that the simple trend of decreasing elevation of the shoreline in a north-westerly direction, illustrated by the linear surface (McCann, 1966), is complicated by a zone of relatively high elevations in Skye, revealed in particular by the 21-ft contour for the surface. The pattern of uniform decrease in elevation westwards and north-westwards, shown by the 45–25-ft contours for the surface, holds true only for the south-western part of the area shown on the map. It may be suggested that the presence of the Post-glacial shoreline in Skye at elevations somewhat greater than might be expected from considerations of the overall pattern of deformation is a result of the greater isotatic depression of the land surface in that area due to the presence, at various stages in the glacial period, of a local Cuillins ice-cap. The trend surface offers no indication that one is dealing with more than one shoreline, and the map of the residuals from the surface, which may on occasion provide additional information, appears to show little of significance in this case.

REFERENCES

CHORLEY, R. J., and HAGGETT, P. (1965) *Trans. Inst. Br. Geogr.*, XXXVII 47.
McCANN, S. B. (1966) *Trans. Inst. Br. Geogr.*, XXXIX 87.
WHITTEN, E. H. T. (1963) *U.S. Office Naval Res.*, Geogr. Branch, Tech. Rep. No. 2.

3 Marking Beach Materials for Tracing Experiments

C. KIDSON AND A. P. CARR

SYNOPSIS

Because of the need to obtain quantitative data on the nature and speed of sediment transport a number of techniques have been devised to mark beach material. Whereas the use of radioactive isotopes and fluorescent tracers has provided new methods, the development of improved resins, dyes and pigments has meant that older techniques remain of value for some purposes.

It is considered that the one method that is suitable for marking and tracing all sizes of beach material and that permits direct tracing both onshore and offshore is that using selected radioactive isotopes. However, because of relatively high cost and fear of health hazards, this technique cannot always be considered. Those techniques that are presently (1962) available are considered individually herein together with the comparative advantages of each.

INTRODUCTION

ONE of the greatest practical difficulties in studying coastal change is to follow the pattern of movement of beach materials both alongshore and offshore. Only by examining the movement of silt, sand or shingle (defined herein as material coarser than granules in the Wentworth scale) is it possible to begin to recognise the fundamental processes that affect the coast. All methods of study depend on some method of marking beach material and on following its movement after injection. The use of radioactive and fluorescent tracers has brought greater precision to coastal studies. In addition, established marking methods have been improved by the use of better materials, such as synthetic resins, while at the same time skin diving has made possible direct observation on the sea bed. It is the purpose of this study to review the various techniques that are now (1962) available.

Tracing experiments can be used both for fundamental studies of coastal change and also for solving practical problems such as the extent of silting and the need for dredging or of coastal defence works.

Until approximately 1954 such features as sea walls and groynes were frequently constructed on an *ad hoc* basis, or were based solely on hydraulic models that were subject to scale effects (Inman and Nasu, 1956). With suitable labelling, rates of erosion, the longshore travel of material and the extent of movement, if any, between the beach and the offshore zone can now all be examined in the field. Such field experiments form an ever more important part of design studies. Tracing experiments are not new. Lord Montagu's work is quoted in the reports of the Royal Commission on Coast Erosion (1907), and N. M. Richardson had conducted experiments at Chesil Bank, Dorset, England, as early as 1902. Nevertheless the techniques available have been considerably improved since 1954.

The method of marking depends on a variety of factors. Among the more important are the size and type of beach material involved, the intensity of beach processes in the experimental area, and the purpose, duration and complexity of the tests. Cost may also be a deciding factor.

Consideration will be given herein to the following: labelling with paints and resins; use of radioactive isotopes and fluorescent dyes; a brief examination of a number of miscellaneous techniques; and finally an appreciation of the comparative value of the various available methods. Tables 3.1 to 3.4 show many of the methods that have been used in marking different types of beach material and also give a summary of a number of the salient features associated with each. (The tables do not attempt to include all the instances of any particular type.)

PAINTS AND RESINS

Perhaps the simplest and the most economical methods for marking beach material are those that rely on paints and resins (Table 3.1). Both types of marking can sometimes be used in conjunction with fluorescent or other dyes as well as pigments. Unfortunately labelling with paint is entirely satisfactory only on the larger fractions of the beach material, especially shingle. With sand and silt, for example, a coating is likely to coagulate a great number of the particles, thereby producing atypical size gradings. Moreover any film on the surface of the material where the ratio of surface area to volume is high represents a fundamental change in the specific gravity and general characteristics. In addition, silt may contain as much as 35 per cent of organic constituents (Allen and Grindley, 1957) and is therefore unsuited for the complete drying that is essential both prior to and

TABLE 3.1

Paints and resins used for marking beach material

Beach fraction[1,2]	Nature of labelling	Remarks	Date of experiment	Site	References
P N	Paint on flint	Not adequately resistant to abrasion	1905	Beaulieu R., England	Royal Commission on Coast Erosion, 1907
P N	Paint (?)	—	1951	U.S.S.R.	Zenkovich, 1956
P A	Marine paint on limestone and artificial tallies	Suffered in abrasion of limestone; better on artificial material	1955 onwards	Bridgwater Bay, England	Kidson and Carr, 1961
P N	Marine paint on flint	Satisfactory for short-term work in severe conditions	1959	Orfordness, England	Kidson and Carr, 1959
P N	Marine paint (with added fluorescent dyes) on flint	Survived continuous immersion for 11 months	1960–1	Scolt Head, England	Kidson et al., 1962
P N	Epoxy resin on flint	Greater resistance than paint, but precise choice of resin	1959–60	Mainly laboratory trials with limited field testing	present paper

[1] P = Shingle; N = Natural; A = Artificial.   [2] Mainly restricted to larger beach fraction.

subsequent to coating. Although, for example, it is possible to label sand with the epoxy resin, crushing is necessary after curing for the requisite time. This may destroy the original grain structure and result in an uneven coating. Paint and resin finishes will therefore be considered only in relation to larger beach material such as shingle.

The most important consideration in selecting a particular finish is durability. The increase in hardness as paint dries is usually governed by the choice of the volatile constituents (Drinberg *et al.*, 1960), although occasionally it is achieved through chemical reaction (for example, catalysed polyurethane). The mechanical properties, such as resistance to abrasion, are determined by the chemical structure and other considerations such as the particular pigments incorporated (Drinberg *et al.*, 1960). Physical strength and chemical stability of the pigments when exposed to the atmosphere or to sea water, and the adhesion of the coating, are important. Some paint films show a sensitivity to water indicated by dullness and blistering as a consequence of swelling, solubility and chemical reaction. Marine paints are usually better than ordinary finishes for the marking of shingle because problems of attack by salt water and alkalis are reduced to a minimum. Developments since the Second World War in both paint ingredients and synthetic resins have improved their durability. Tests at Orfordness, Suffolk, England (Kidson and Carr, 1959), under severe conditions showed that most modern marine paints had adequate resistance for short-term work. Even in experiments off Scolt Head Island, Norfolk, England (Kidson *et al.*, 1962), where pebbles were continuously immersed in sea water for periods of as much as eleven months, only minor deterioration occurred in the finish.

These field experiments have been supplemented by laboratory tests conducted by the Coastal Research Section of the Nature Conservancy. The purpose of laboratory investigation has been to subject a variety of finishes to simulated beach conditions, including immersion and agitation in salt water and abrasive contact with one another, and in this manner to evaluate the relative suitability of different materials. Flint pebbles were used for this work because of their hardness. This enabled the loss of the paint or resin coating to be measured by weighing before and after an experiment. Other beach materials such as Liassic limestone could lose a significant proportion of their original weight through abrasion during the tests. The highest resistance to abrasion was found with an oven-cured epoxy resin. The coating was applied by dipping the pebbles into the resin and then draining them on wire mesh. Curing followed for 160 minutes at 180°C., as directed by the manufacturers (C.I.B.A., 1960). It was an advantage to practice de-greasing and to have the resin somewhat thinner than recommended for normal use in order to avoid bubbles being trapped in the surface coating as it hardened, and to improve adhesion. Best results appear to be obtained when the temperature is gradually brought up to 180°C. from cold. Only when bubbles occurred was durability at all suspect. Where pigments (for example, chromic oxide) or dyes (fluorescent or otherwise) were incorporated, such changes as took place were largely a reflection, not of the resin itself, but of the colouring matter. Other synthetic resins were less satisfactory even when it was possible to effect curing at room temperature. Mineral-filled constituents, a colour cast in the initial mixture or difficulties in application all posed problems.

Experiments were also conducted in order to discover, in the case of conventional paints, the extent to which heat drying affected durability. The paint was applied in a manner similar to that used for the epoxy resin. Usually only one coat was given, although a second coat can often be of advantage. Finishing, rather than under-coats, were always used. Artificial drying reduces the tendency to blister in some paints and is essential in any case in which large quantities of marked material are needed; otherwise handling time becomes excessive. Nevertheless heat drying may reduce the flexibility of the coating by making it harder but more brittle.

Satisfactory results with heat drying depend on resistance both on the part of the pigments and the base to the effect of temperature (Gardner and Sward, 1946; Remington and Francis, 1954). In the broad sense the writers' work showed that most of the bases tried, whether of a linseed-alkyd or an ester composition, performed satisfactorily at the temperatures and over the time used (30 minutes at 100°C.). Pigments used were more critical. Thus in one range of paints both yellow and pink retained their colour satisfactorily, whereas orange did not.

Catalysed polyurethane finishes were also examined. They gave a performance better than conventional marine paints, but below that of the best epoxy resin. Polyurethane hardens fairly quickly by chemical reaction, but a long curing time is necessary afterwards. This may be reduced by the application of heat.

High light reflectance is also of importance, specially where diving for samples is carried out underwater. Some shades of yellow, for example, can reflect more than half the light obtained in daylight (Gardner and Sward, 1946). In the underwater experiments at Scolt Head Island, in a depth of 15–20 ft, one shade of pink was found particularly satisfactory for identification, even in the poor visibility prevailing at the time. The same bright colours above the water line are, however, likely to attract unwelcome attention from casual visitors on beaches where public access is a problem.

Marine paint and synthetic resin finishes provide a cheap and fairly satisfactory method of marking the larger fractions of beach material. In most cases heat processing is essential, except where hardening is achieved through chemical reaction rather than solvent evaporation. Even then reduction of curing times remains an advantage.

RADIOACTIVE TRACERS

General considerations

With radioactive tracers (Table 3.2) it is possible to detect marked beach material both on the beach and on the sea bed, and, with suitable detection equipment, even to locate markers when buried in the beach (Kidson *et al.*, 1958). This is the only method that is applicable to all sizes of material. Field experiments using this form of tracing were first carried out in Japan (Inose and Shiraishi, 1956) and England (Putman and Smith, 1956) in 1954. The technique has been applied in many countries for marking silt, sand and shingle. Both artificial and natural beach material have been labelled. Surface labelling has been carried out on natural as well as artificial material, whereas mass labelling has been done only on artificial material. Amounts of activity have varied within wide limits (Greisseier, 1960). A large number of isotopes are suitable and many have been used. The value of any particular isotope is governed by various considerations. The half-life of the isotope must be appropriate to a particular experiment. For example, in order to examine the effects of an individual storm, a tide or a tidal cycle, or to enable repeat experiments at given intervals of time, a short half-life would be required. This also minimises health hazards. A capacity to take up the necessary specific activity and ease in handling are other considerations. The choice of element is also influenced by certain physical and chemical properties relating to the type and intensity of rays emitted, fusion, adsorption, reduction, ion-change and others. A large number of papers have been written describing both the isotopes used and the equipment available. Particularly comprehensive references will be found in the work of Greisseier (1960).

A tracer must fulfil a number of criteria. It is essential that it should behave as the material being traced. It should also remain incorporated in or on the material in order that changes in recorded values reflect only dispersion of the material and natural decay; there should be no question of mechanical change or chemical reaction. Putman and Smith (1956) considered, in the case of sand, that where the intensity was being directly measured, the initial activity should be proportional to mass. This necessitates bulk labelling rather than labelling of the surface layer only because false results would be obtained if differential sorting of sizes took place. The quantity of marked sand must be sufficiently great that, even where dispersal occurs over a large area, inconsistent readings resulting from erratic distribution of single particles will be avoided. These conditions apply equally well to silt. In the case of shingle tracing, however, where individual contacts are recorded, mass labelling is not required.

Experiments with labelled sand and silt

Work with radioactive tracers was initially concentrated on sand and silt. Where direct detection was used, it was often the practice in Great Britain to use soda glass into which a suitable element had been incorporated. This was generally scandium and usually in the form of its oxide. After grinding to reduce the glass to the appropriate size range, the tracer was placed in a reactor for irradiation (Greisseier, 1960; Putman and Smith, 1956). This produced Sc^{46} which has a half-life of 85 days. The same method has been used in France (Hours and Jaffry, 1959). In France (Jaffry and Hours, 1959) and elsewhere, however, notably in Japan (Inose and Shiraishi, 1956) and the Netherlands (Arlman *et al.*, 1957, 1960), radioactivity has generally been incorporated prior to grinding. In these circumstances the activity is necessarily limited in intensity because of the need for protection of personnel in the subsequent pulverisation process. In the initial Japanese experiments Zn^{65} was used, but because of its low gamma-ray emission scandium was considered more suitable for later work. Ta^{182} has been tried in France.

A number of isotopes have been used to label the natural beach material through adsorption of chemical solutions on the sand surfaces. This adsorption is increased by appropriate treatment such as annealing. Relatively large quantities can be treated, but the method suffers from variations in adsorption rates between grains

TA

Radioactive isotopes use

Beach fraction[1]		Nature of labelling				
					Rays emitted	
		Labelling remarks	Isotope	Type	Principle (MeV)	Half- (day
						(a)
M	A	Mass labelling after grinding of soda glass				
M	N	Applied as solution				
S	A	Mass labelling after grinding of soda glass	Sc⁴⁶	γ	0·89 1·12	85
S	A	Applied to Zeolite 'green-sand'				
S	A	Isotope added to soda glass	Zn⁶⁵	γ	1·11	245
S	A	Glass beads	Na²⁴	Y	1·37 2·76	0·6
S	N	Surface labelling				
S	A	Crushing of soda glass subsequent to labelling	Cr⁵¹	γ	0·32	28
M	N	Applied as solution	Au¹⁹⁸	γ	0·41	2·7
S	A	Mass labelling in soda glass	Ru⁸⁶	β γ	1·77 1·08	18·6
S	N	Agar-agar film labelling	Fe⁵⁹	γ	1·10 1·29	45·1
S	N	Radiation of phosphorus impurity in quartz sand	Ph³²	β	1·71	14·3

(continued on pp. 78–9)

[1] M = Silt; S = Sand; P = Shingle; A = Artificial; N = Natural.

king beach material

Remarks	Date of Experiment	Site	References

Silt

Remarks	Date of Experiment	Site	References
Represents mineral fraction only. Artificial tracer; quantitative.	1954, 1955	Thames estuary	Putman and Smith, 1956 Allen and Grindley, 1957
Affects all silt components, mitigates flocculation problems	—	—	Krone, 1960
Artificial tracers; quantitative results	1954–7	Poole, England; Great Yarmouth, England; Adour, France	Putman and Smith, 1956 Reid, 1958 Jaffry and Hours, 1959
Close approx. to natural materials	1958–9	Netherlands Delta Project	Artman et al., 1960
Greater control of specific gravity but lower specific activity available	1954, 1955	Hokkaido, Japan	Inose and Shiraishi, 1956
Short half-life; glass beads atypical. Pilot experiment only	1956	Netherlands	Artman et al., 1957, 1960
Variable adsorption with natural material	1957	Baltic, Sweden	Davidson, 1958 Greisseier, 1960
Difficulties in direct detection with low γ emission	1955	La Bocca, France	Hours et al., 1955
Scintillation counter-detection through low γ. This is only field-tested method so far of labelling non-mineral fraction in mud	1958–9	San Francisco, U.S.A.	Krone, 1958, 1960
Low percentage γ disintegrations and adsorption of β-rays, rejected after pilot trials	1955	France	Hours et al., 1955
—	1956	U.S.S.R.	Afanasev et al., 1957
Indirect detection requiring sampling. Limited application. Much laboratory work involved.	1955 onwards	California, U.S.A.	Inman and Chamberlain, 1959

TABLE 3

Beach fraction[1]		Nature of labelling				
					Rays emitted	
		Labelling remarks	Isotope	Type	Principle (MeV)	Half-l (day
						(a) S
S	N	Reduction of silver nitrate solution	Ag^{110}	β	0·09 0·53	270
S	A	Incorporated into soda glass	Ta^{182}	γ	Various	111
						(b)
P	—	Activity placed in hole in pebble	Ta^{182}	γ	Various	111
P	A	Activity placed in resin in hole in cement marker				
P	N	Activity adsorbed on to surface	Ba^{140} $La^{140}-$	— γ	— 1·60	12·8 2·2
P	N	Activity adsorbed on to surface	La^{140}	γ	1·60	2·2

[1] M = Silt; S = Sand; P = Shingle; A = Artificial; N = Natural.

composed of different minerals and may provide problems of abrasion and solution over a long period. It also precludes the measurement of intensity of radiation. Agar-agar has been used for fixing radioactive Fe^{59} (Afanasev *et al.*, 1957). Cr^{51} has also been adsorped on to sand (Davidson, 1958). Cr^{51} is not really a satisfactory tracer because of its low gamma-ray emission and poor penetration. Nevertheless, together with Au^{198} and Ir^{192} it is valuable for studying problems in which a health hazard arises. Ag^{110} has also been used for surface coating in experiments in Portugal (Gibert *et al.*, 1958), in the form of silver nitrate to which a reducing agent was added. Sand was wetted and exposed to sunlight which, by reducing the silver salt, produced metallic silver on the grains. This is stated to have withstood agitation quite satisfactorily. With this method it is supposedly possible to use large quantities of labelled sand without *inued from pp. 76–8*

Remarks	Date of Experiment	Site	References
Silt			
Requires sampling. Silver recovered chemically in laboratory. Large quantities of tracing materials can be employed. Variable adsorption.	1957	Cape Mondego, Portugal	Gilbert *et al.*, 1958
Mass labelling	1957	Adour, France	Jaffry and Hours, 1959
Shingle			
Individual contacts recorded. Preparation of markers tedious	1956	Var, France	Jaffry and Hours, 1959
Easy handling. Individual contacts recorded. Preparation of markers tedious	1956	Scolt Head, England	Kidson *et al.*, 1956
Easy handling. Individual contacts recorded. Requires material resistant to surface abrasion	1957 and 1959	Orfordness, England	Kidson *et al.*, 1958 Kidson and Carr, 1959
Individual contacts recorded. Requires resistant material	1959	Orfordness, England	Kidson and Carr, 1959

exceeding permissible levels of activity. However, greater quantities of the chosen isotope must be used in order to obtain the same sensitivity, and this may well give rise to similar health hazards. In addition, samples must be taken because direct readings are not feasible. This is followed by chemical extraction of the silver or detection with photographic plates.

Another experiment using the natural beach material was conducted in the United States by Inman and Chamberlain (1959). Quartz sand that contained phosphorus was irradiated. Most natural sand is unsuited for irradiation as the only isotopes that are produced are short-lived, being mainly Si^{31} and Na^{24} with half-lives of 2·6 and 15·0 hours respectively. Even Ph^{32} is not ideal. Its greatest disadvantage, however, is the impossibility of using direct tracing methods, since only beta rays are emitted, although injection is simplified because it can be carried out by skin divers. Where gamma rays are produced and direct tracing methods are used, injection devices have had to be designed to protect the operatives and to prevent dispersal of the tracer before it reaches its specified depth or the sea bed.

In the experiments related to the Netherlands Delta Project, initial trials were carried out with short-lived Na^{24} and glass beads. The latter were found unsatisfactory because of their higher specific gravity and greater size than the natural material. Later work was concentrated on Sc^{46}. Because it was found impossible to attach this directly to the sand surface, the isotope was bonded to an inorganic ion-exchange substance. This zeolite 'greensand' had a specific gravity close to North Sea sand (2·72 to 2·76 as opposed to 2·65 to 2·68) and can readily absorb the necessary scandium. This feature could not be used to its full advantage because of health hazards and statistical considerations. Unfortunately the material is soft and therefore readily abraded, thus reducing the grain size from that typical of the natural beach material. The defect can, however, be remedied by firing the 'greensand' in order to harden it.

Silt movement has been treated in a manner similar to the treatment of sand. The investigations (Putman and Smith, 1956) in the Thames estuary in 1954 relied on ground glass incorporating the tracer Sc^{46}. Even at best, this type of tracer can only represent the mineral fraction in mud; the method has been subject to criticism by Krone (1958) and Jaffry and Hours (1959). The latter noted that mud is partly organic and largely fibrous, with only small mineral particles. Nevertheless, in the tracing experiments in the Thames estuary steps were taken to ensure that this problem was minimised. By pre-mixing with a large volume of real mud, it was possible to ensure that the mineral fraction was enmeshed in the fibrous material and did behave like the mass of the mud. This technique avoided drying the actual mud and merely involved extraction of the mineral sample to which the ground glass is matched. More recently, Krone (1958) has used gold in the form of a solution. He considered Sc^{46} suitable for long-term work if applied in a similar manner. The structure and composition of mud has a direct bearing on the way in which most injections have been conducted. Irradiated material has usually been mixed with the natural mud prior to injection in order to account for these points. It would appear that a number of the results so far obtained in experimental work have been considered by some workers as not entirely valid.

Radioactive labelling also has possibilities in accretion experiments. For example, the extent of build-up of sand dunes could be followed by the progressive attentuation of gamma rays from a buried Co^{60} source.

Experiments with labelled shingle

Experiments for tracing pebbles have been fewer in number, but again early tests were dependent on artificial material.

In the first experiments in England, off Scolt Head Island, Norfolk (Kidson *et al.*, 1956), in 1956, Ba^{140}–La^{140} was used. This tracer was inserted into holes made in concrete markers and then sealed in resin. Barium has a half-life of 12·8 days, which is convenient for experiments of moderate duration. Carrier-free Ba^{140} is separated by chemical processing from its daughter product La^{140}, thus enabling labelling and handling to be carried out more safely with the low-energy gamma emitter before the build-up of the strongly gamma-emitting La^{140}. The activity builds up to a maximum approximately three days after the initial separation. A similar form of labelling with pebbles was undertaken, in France, later in 1956 using Ta^{182} (Jaffry and Hours, 1959).

In the following year, drift experiments lasting for two months at Orfordness (Kidson *et al.*, 1958) again used Ba^{140}–La^{140}. Improved techniques made it possible to absorb the activity on to the natural flint material. For shortlived experiments over part of a tidal cycle, La^{140} has been used alone (Kidson and Carr, 1959). This has a half-life of 40 hours.

Analysis and plotting

Methods of analysing and plotting the results are largely governed by the method of tracing and the type of material labelled. Accuracy of measurement is determined by the statistical pattern of the radio-active transformation processes, the quantity and activity of isotopes used and the measuring conditions under which detection is carried out. This in turn is affected by the type and efficiency of the detecting apparatus and the time available for searching or sampling. A broad general distinction may be made between isotopes that emit gamma rays and those emitting beta rays. With gamma rays from suitably chosen isotopes producing emissions of approximately 1 mega-electron volt (1 MeV) or above, detection is direct. Geiger-counters or scintillation-counters may be towed over the beach or the sea bed and contacts recorded.

Where beta rays are produced, indirect methods are necessary.

Sampling, and laboratory analysis, which may be laborious, must precede plotting. For quantitative results with beta emitters, it is essential that samples be uniform in thickness and that cores be of constant length. In one series of experiments (Inman and Chamberlain, 1959), divers obtained samples that were analysed by autoradiographic means with X-ray films. In this same experiment it was found convenient to erect a grid on the sea bed in order to facilitate sampling. A second distinction is necessary between surface and mass labelling. Where material is surface-labelled, individual contacts (not necessarily individual markers, however) are recorded.

Mass labelling makes possible the measurement of the intensity of radioactivity, and isopleths, corrected for decay, can be drawn. In all cases conventional survey methods, such as the use of horizontal sextant angles or theodolite intersections, can be used for locating and plotting. Techniques of activity measurement are dealt with at greater length by Arlman *et al.* (1957), Jaffry and Hours (1959), and Krone (1958).

Safety considerations

Precautions are necessary to minimise hazards to health resulting from the use of isotopes. There are two aspects to this problem. First, the scientists conducting the experiments come into relatively close contact during both the preparation and the injection of the irradiated material and adequate precautions are necessary at this stage; second, the general public may have access to the experimental area or may come into contact with marked material that escapes from that area. It is important to ensure that the level of activity is sufficiently low so that no danger to health exists. The greatest dangers are in the actual preparation of the material rather than in its use. Careful choice of isotopes with a suitable half-life and nature and intensity of ray emission greatly aid in solving most difficulties. Care should also be taken in the transport of the isotope and in its injection to avoid contamination of the working area. Protection is obtained through distance from the source; by absorption of the rays, as with lead or concrete shields; and by protective clothing. Dosimeters can give a check on the actual amount of activity received. Warning notices may be advisable. If storm conditions prevent injection or impose a delay in injection, great care is necessary in the storage of irradiated material. There is a tendency by the general public to exaggerate health hazards. In most experimental work the level of activity is so low that for practical purposes no such hazard exists.

Radioactive labelling has provided a useful tool for research into beach movement. All fractions of beach material have been marked, even if some qualifications arise as to the labelling of silt and mud. However, the greatest value in the method is the direct results that can be obtained by using isotopes with suitable gamma-ray emission. There would appear to be little point in using radioactive tracers if samples are going to be taken. If samples are essential, marking with fluorescent dyes may well be of greater value.

FLUORESCENT TRACERS

The use of fluorescent dyes (Table 3.3) for following the movement of beach material was pioneered by Russian scientists (Zenkovich, 1958), although more recently the technique has also been applied in Great Britain (Reid and Jolliffe, 1961). Work in England has not yet (as of 1962) been so conclusive (Hydraulics Research Board, 1959, 1960) as that in the Soviet Union. V. P. Zenkovich and others carried out experiments with sand labelled with a hydrophile colloid that contained the fluorescent material. The binding substances consisted of agar-agar, bone glue, gum and starch. The agar–bone glue mixture is stated to have retained the dye for approximately three months before gradually dissolving in water. With this method Zenkovich claimed it was possible to detect one grain in 10 million.

The choice of the fluorescent substance is restricted by such factors as decomposition in sunlight or by the dye's properties in attacking skin, as in the case of anthracene. The quantity of dye used depends on the chemical itself, the size of the grains to be coated and, hence, the surface area and the natural colour of the beach material. Where samples are taken, they can be examined by means of luminoscopes on the beach or any ultra-violet light source.

There are a number of disadvantages with the technique, however. Pre-eminent among these is the necessity to collect samples, even though these may be analysed on the site, or, alternatively, the need to search the beach above the water line in the dark with fluorescent lamps at low water. Underwater samples must be obtained by cable ropeways or from boats with a Van Veen grab. Alternatively, skin divers can be employed either to collect material for subsequent analysis or for carrying out searches with waterproofed ultra-violet lamps. In addition to the normal difficulties of examining the sea bed

TABLE 3.3

Fluorescent dyes used for marking beach material

Beach fraction[1]	Nature of labelling	Need for samples or detection at night	Date of experiment	Site	References
S N	Anthracene and 'lumogene' in agar–bone glue base	Anthracene difficult to handle. Detection claimed of 1 grain in 10^7	1953 onwards	Black Sea, U.S.S.R., and elsewhere	Zenkovich, 1958
S N	Anthracene and rhodamine	Anthracene unstable. Loss of tracer.	1958 and 1959	Dawlish Warren, Devon, England	Hydraulics Research Board, 1959, 1960
P A	Various dyes incorporated in concrete–dolerite aggregate	Searching done at low tide during darkness. Problems of fracturing and rate of abrasion	1950–60	Kent, England	Reid and Jolliffe, 1961
P N	Incorporated in epoxy–resin coating	Choice of dye depends on compatibility with resin and stability on heating during curing process	1959–61	Mainly laboratory trials with limited field testing	See present paper

[1] P = Shingle; S = Sand; N = Natural; A = Artificial.

with divers, the use of ultra-violet lamps introduces other problems. The work is usually performed at night when the fixing of contacts is less easy. Cables from a main source of electricity or a portable generator restrict the area that can be searched at any one time. Some self-contained ultra-violet units have been used (Woodbridge and Woodbridge, 1959) and satisfactory results have been obtained by using divers at night.

In Great Britain, Reid and Jolliffe (1961) have conducted field experiments with both pebbles and sand. In their tests with artificial markers on shingle beaches, the particles of resin in which the fluorescent dye had been incorporated were added to a cement aggregate that included quartz dolerite. The fragments of quartz dolerite made it possible to produce a cement pebble of a specific gravity comparable to that of the beach shingle. This technique may, however, produce a form of marker that is initially angular, liable to fragmentation and also subject to a rate of abrasion not necessarily similar to that of the shingle on the beach. Where the natural beach material is sufficiently resistant, it is possible to avoid these difficulties.

Mention has been made of the practicability of incorporating fluorescent dyes in surface coatings of paints and resins. In laboratory tests a number of fluorescent dyes were added to an epoxy resin and then cured at 180°C. Because it is essential that the surface layer be moderately thick in order to keep adequate fluorescing properties, this factor again restricts the technique to shingle. In addition the dye must dissolve in the solvents used with the resin and must also be unaffected by the temperature necessary for curing. Of the fluorescent substances tried (among them two anthracenes, rhodamines B and 6G, and fluorescein sodium), only fluorescein sodium was completely satisfactory. It produced a conspicuous uniform yellow coating. No doubt further research would provide other suitable indicators. Even this method is subject to some of the limitations encountered by Reid and Jolliffe. It is still necessary to search the beach with an ultra-violet lamp and portable generator at low water during the hours of darkness, as the collection of samples is clearly impracticable with shingle. This in turn restricts the area that can be examined because of the limited time available between tides.

Reid and Jolliffe noted the that use of two different wave-lengths of light makes possible the identification of as many as eight different fluorescent substances. Among those cited were rhodamine B, primuline and uvitex. This makes the method much more versatile.

TABLE 3.4

Miscellaneous techniques used for marking beach material

Method	Beach fraction[1]	Nature of labelling	Remarks	Date of experiment	Site	Reference
Dyes	S N	Aniline	Difficulties in detection	1947 (?)	U.S.S.R.,	Zenkovich, 1958
	S N	Waxoline	Used in accretion experiments; not light-fast	1957 (?)	U.S.A.	Inman and Chamberlain, 1959
				1955 and 1958 onwards	Braunton Burrows, Devon, England	Kidson and Carr, 1960
	P N	Engineering dyes under proprietary names	Thin coating; not sufficiently dense or resistant	1956–8	Norfolk, England	Harding, 1961
Inorganic coatings	S N	Sodium silicate and chromic oxide	Baking problems including difficulties of coagulation	1954–5	Laboratory trials	—
Coloured glass	S A	Crushed stained glass	Artificial tracer. Difficult to detect	1939, 1944, 1955	Germany, France (with radioactivity)	Geib, 1944, Wasmund, 1939, Hours

Method	Material[1]	Type	Description	Comments	Date	Location	Reference
and metal markers			material broken to beach shingle sizes	angularity, special gravity, abrasion characteristics	onwards	England, Bridgwater Bay, England	and Carr, 1961
Wire tags	P	A	Incorporated into or through holes in artificial tallies	Enables wide range of categories to be used	1955 onwards	Bridgwater Bay, England	Kidson and Carr, 1961
Etching	P	N	Acid into limestone pebbles	Time-consuming. Tracing difficulties			See, for example, Kidson and Carr, 1964
'Foreign' rocks and materials	PS	E	Injection of minerals or absent from area	Problems of specific gravity. Painstaking analysis with sand	Various	e.g. Devon and Cornwall, England, and U.S.S.R.	Stuart and Simpson, 1937, Steers, 1960, Zenkovich, 1956
Coal dust-sand artificial horizon	M	E	Use of 'alien' layer to study vertical build-up (e.g. in salt marshes)	Assumes no consolidation below artificial horizon	Various	e.g. Scolt Head Island, England	Reid, 1958

[1] P = Shingle; S = Sand; M = Silt; N = Natural; A = Artificial; E = Natural but not indigenous.

Clearly, the use of fluorescent marking has important potentialities, but it also poses sizeable problems.

MISCELLANEOUS TECHNIQUES

Apart from the major methods previously considered, a number of other means of labelling have been tried (Table 3.4). Richardson (1902) used brickbats at Chesil Bank, Dorset. In experiments to study the rate of shingle movement at Bridgwater Bay, Somerset, fireclay markers were used (Kidson and Carr, 1961). The latter were chosen to conform closely to the physical properties of the sandstone and limestone beach material, which could not be used directly because of its abrasion characteristics.

Large material, from coarse sand upwards, can be surface-coated in a variety of other ways also using pigments or dyes. Sand, for example, can be coloured by the use of chromic oxide in a sodium silicate solution that is then subjected to baking. It is also possible to label the same material with waxoline dyes, although these are subject to fading under prolonged exposure to light. They have been used in accretion experiments at Braunton Burrows, Devon (Kidson and Carr, 1960). Russian scientists and others have tried aniline dyes for marking sand (Zenkovich, 1958). Some of these methods tend, as does the coating of sand with resins, to render the sample of beach material no longer typical because of the high surface area and the resulting high proportion of marking medium to beach material. Thinner liquids with rapid-drying solvents and variable quantities of dissolved or suspended colours have been tried. However, although no longer open to the previous objections, these are not sufficiently dense to resist any appreciable abrasion and soon become difficult to identify. Some were originally developed for engineering purposes but have been applied to coastal problems (Hardy, 1961).

Foreign material, such as coal dust or natural sand containing minerals alien to the particular site under investigation, can be used to study accretion and sediment transport (Stuart and Simpson, 1937). Not only are numerous samples needed for identification in this method of studying sediment movement, but, as Jaffry and Hours have noted, the technique necessitates much laboratory analysis. The method, as well as that practised with a number of other forms of labelling, is open to the objection that experiments are being carried out with atypical material and the results cannot be directly applicable. Coloured glass has also been tried.

Some attempts have been made to etch or drill beach material. This can be done with less resistant rocks, but where movement and abrasion are great, the method is of little value.

A number of instances have been noted of the use of divers in collecting samples, as in experiments off Scripp's Beach, California (Inman and Chamberlain, 1959). Although in no sense a means of marking, diving is relevant to the manner in which tracing methods can be applied under water. The limitations of this form of investigation are considered by Kidson, Steers and Flemming (1962). Both storm conditions and strong currents restrict the use of the method, while at the same time, at least in water around the British Isles, visibility is often too poor for satisfactory observations.

SUMMARY AND CONCLUSIONS

Choice of material for tracing

Theoretically, it is desirable to use the material of the beach for drift experiments, but circumstances may prevent this. For example, the material may be so soft that abrasion can remove any surface coating that is applied to it. Alternatively, the need for a large number of different categories during a single experiment may necessitate recourse to some form of artificial tally such as those used at Bridgwater Bay (Kidson and Carr, 1961). Similarly, with radioactive tracers, if it is intended to follow the intensity of radiation rather than the mere presence of irradiated sand, bulk labelling is required and this in turn means that an artificial tracing material is needed. Soda glass is commonly used. Finally, the difficulty in labelling silt in any manner, and the need for quantitative results, had resulted, until 1957, in the exclusive use of irradiated soda glass rather than the natural material.

In cases in which substitutes for the silt, sand or shingle of the beach must be used, several conditions must be satisfied. Artificial markers must correspond to the actual material as closely as possible in terms of weight, size, shape and specific gravity. The tallies should respond to beach processes in the same manner as the natural material in all respects, and a pilot test to establish this correspondence may be necessary. It is highly undesirable that artificial markers should be more susceptible to fragmentation or abrasion, as this produces size gradings totally different from those of the natural beach constituents. Even where all these qualifications are met, it still remains impossible to state with categorical assurance that the rate of movement of the artificial tallies is a precise record of that which would obtain had it been possible to use the natural beach material.

Choice of form of labelling

The choice of a tracer depends on the nature of the beach material, the site and the purpose of the experiment. The three most usual methods, labelling with fluorescent dyes, paints and radioactive tracers, all possess advantages and disadvantages.

Fluorescent dyes have been used to mark sand and pebbles. Tracing requires numerous samples, a method only really possible with sand beaches, or, alternatively, direct recording on the beach. This entails either the use of suitable filters in bright sunlight or ultra-violet lamps during the hours of darkness. The method tends to be time-consuming and of somewhat limited application.

Paints and resins are of value for labelling the larger fraction of beach material both above and below water. Coated pebbles can be followed underwater by skin divers where visibility is adequate. Work can be carried out during daylight and does not need the complexity of equipment required with the fluorescent method.

However, the most satisfactory means of marking all sizes of beach material, and the only method so far capable of giving any result for following silt movement, is undoubtedly that using radioactive tracers. This method is most useful where gamma rays are detected, as this enables direct recording. This is especially valuable where only short periods of calm weather are likely. Greater areas can be searched and painstaking laboratory analysis eliminated. The higher initial cost may be offset by economies in analysis and interpretation. Above water, irradiated material may be identified even when buried in the beach. Bulk labelling can be utilised in order to interpret radioactivity measurements quantitatively, although in this respect the method becomes open to the criticisms directed against artificial material. However, Krone (1958, 1960) considered that the advantages were overwhelming. Where the natural beach material is resistant, as with flint shingle or quartz sand, surface labelling is possible. This is most suitable for pebbles, where individual contacts are recorded. Problems concerning lack of flexibility, handling and health hazards are significant, but a careful choice of isotope can minimise these. With isotopes emitting beta rays, the need to obtain samples for analysis removes any real advantage over alternative forms of marking. Fluorescent dyes might have greater potentialities in these restricting circumstances.

Other means of labelling have been used, all with varying success, largely dependent on the site and the specific application, as for example where a large number of categories are required for comparative results. However, whatever method is practised, because recovery rates are liable to be low it is highly desirable that the maximum quantity of marked material is used.

In the broad sense the largest fraction of beach material can be marked most satisfactorily because the choice of means is greatest. Sand can be labelled adequately by fluorescent or radioactive means, but silt has remained something of a problem. Early experiments with radioactive tracers, such as those in the Thames estuary, relied on finely ground glass particles to match those of the minerals that occur in mud. However, a large proportion of mud constituents are organic. Labelling with radioactive gold has been the most encouraging method with silt. Since 1950 greater progress has been made in the labelling of beach material and thus in recording the rate and nature of alongshore and offshore movement than at any time previously.

REFERENCES

AFANASEV, V. N., ZVEZDOV, V. N., LEONTEV, O. K., and JAKOVLEV, S. G. (1957) 'Instruction in the use of radioactive isotopes in studying the movement of beach material', *Sojuzmorprojekt*, State Research Inst. Internal Report (Geographical Faculty, Lomonossov Univ., Moscow).

ALLEN, F. H., and GRINDLEY, J. (1957) 'Radioactive tracers in the Thames estuary', *Dock and Harbour Authority*, XXXVII 302–6.

ARLMAN, J. J., SANTEMA, P., and SVASEK, J. N. (1957) *Movement of Bottom Sediment in Coastal Waters by Currents and Waves: Measurements with the Aid of Radioactive Tracers in the Netherlands* (Ministry of Transport and Waterways, The Hague).

——, —— and VERKERK, B. (1960) 'The use of radioactive isotopes for the study of littoral drift', *Philips Technical Review*, XXI 157–66.

C.I.B.A. (A.R.L.) LTD (1960) *Araldite 985-E*, Instruction Sheet No. S1 (Duxford, Cambs.).

DAVIDSON, J. (1958) 'Investigations of sand movements using radioactive sand', *Lund Studies in Geography*, series A, *Physical Geography*, XXII 107–26.

DRINBERG, A. YA., GUREVICH, E. S., and TCKHOMIROV, A. V. (1960) *Technology of Non-metallic Coatings* (Pergamon Press, London).

GARDNER, H. A., and SWARD, G. G. (1946) *Physical and Chemical Examination of Paints, Varnishes, Lacquers and Colors*, 10th ed. (H. A. Gardner, Bethesda, Md).

GEIB, K. (1944) 'Meeresgeologische Untersuchungen im Bereiche der Ostpommer', *Geol. d. Meere und Binnengewässer*, VIII.

GIBERT, A., *et al.* (1958) 'Tracing sand movement under sea water with radio-active silver Ag¹¹⁰, *Proceedings*, 2nd U.N. Internatl. Conf. on Peaceful Uses of Atomic Energy, XIX 355–9.

GREISSEIER, H. (1960) 'Zur Anwendung von Radioisotopen beim Studium der litoralen Materialbewegungen', *Acta Hydrophysica*, VI 163–86.

HARDY, J. (1961) 'The movement of beach material near Blakeney Point, Norfolk' British Association Meeting, Section E (Norwich).

HOURS, R., and JAFFRY, P. (1959) 'Applications des isotopes radioactifs à l'étude des mouvements des sédiments et des galets dans les cours d'eau et en mer', *La Houille Blanche*, XIV 318–47.

——, NESTEROFF, W. D., and ROMANOVSKY, V. (1955) 'Méthode d'étude de l'évolution des plages par les traceurs radioactifs', *Travaux du Centre de Recherches et d'Études Océanographiques*, XI 1–7.

HYDRAULICS RESEARCH BOARD (1959) *Hydraulics Research 1958* (Dept of Scientific and Industrial Research) pp. 63–4.

—— (1960) *Hydraulics Research 1959* (Dept of Scientific and Industrial Research) pp. 55–6.

INMAN, D. L., and CHAMBERLAIN, T. K. (1959) 'Tracing beach sand movement with irradiated quartz', *Journal of Geophysical Research*, LXIV 41–7.

—— and NASU, N. (1956) 'Orbital velocity associated with wave action near the breaker zone', *Technical Memorandum No. 79*, Beach Erosion Board (U.S. Dept of the Army, Washington, D.C.).

INOSE, S., and SHIRAISHI, N. (1956) 'The measurement of littoral drift by radio-isotopes', *Dock and Harbour Authority*, XXXVI 284–8.

JAFFRY, P., and HOURS, R. (1959) 'L'Étude du transport littoral par la méthode des traceurs radioactifs', *Cahiers Océangraphiques*, XI 475–98.

KIDSON, C., and CARR, A. P. (1959) 'The movement of shingle over the sea bed close inshore', *Geogr. J.*, CXXV 380–9.

—— and —— (1960) 'Dune reclamation at Braunton Burrows, Devon', *Chartered Surveyor*, XCIII 298–303.

—— and —— (1961) 'Shingle drift experiments at Bridgwater Bay, Somerset', *Proc. Bristol Nat. Soc.*, XXX 163–80.

——, —— and SMITH, D. B. (1958) 'Further experiments using radioactive methods to detect the movement of shingle over the sea bed and alongshore', *Geogr. J.*, CXXIV 210–18.

——, STEERS, J. A., and FLEMMING, N. C. (1962) 'A trial of the potential value of aqualung diving in coastal physiography on British coasts', *Geogr. J.*, CXXVIII 49–53.

——, —— and SMITH, D. B. (1956) 'Drift experiments with radioactive pebbles' *Nature*, CLXXVIII 257.

KRONE, R. B. (1958) *Silt Transport Studies Utilising Radioisotopes*: *First Annual Report, October 1956–December 1957* (Univ. of California Inst. of Engrg. Research, Berkeley, Calif.).

—— (1960) *Methods for Studying Estuarial Sediment Transport Processes* (Univ. of California Inst. of Engrg. Research, Berkeley, Calif.).

PUTMAN, J. L., and SMITH, D. B. (1956) 'Radioactive tracer techniques for sand and silt under water', *Internatl. J. Appl. Radiation and Isotopes*, I 24–32.

REID, W. J. (1958) 'Coastal experiments with radioactive tracers', *Dock and Harbour Authority*, XXXIX 84.

—— and JOLLIFFE, I. P. (1961) 'Coastal experiments with fluorescent tracers', *Dock and Harbour Authority*, XLI 341–5.

REMINGTON, J. S., and FRANCIS, S. (1954) *Pigments* (Leonard Hill, London).

RICHARDSON, N. M. (1902) 'An experiment on the movements of a load of brickbats deposited on Chesil Beach', *Proc. Dorset Nat. Hist. Field Club*, XXIII 123–33.

ROYAL COMMISSION ON COAST EROSION (1907) *Evidence of Lord Montagu*, I (2) 365.

STEERS, J. A. (ed.) (1960) *Scolt Head Island*, 2nd ed. (Heffer, Cambridge).

STUART, A., and SIMPSON, B. (1937) 'The shore sands of Cornwall and Devon from Land's End to the Taw–Torridge estuary', *Trans. Roy. Geol. Soc. Cornwall*, XVII 13–14.

WASMUND, E. (1939) 'Färbung und Glaszusatz als Messmethode marines Sand und "Geröllwanderung" ', *Geol. d. Meere und Binnengewässer*, III.

WOODBRIDGE, R. G., and WOODBRIDGE, R. C. (1959) 'Application of ultra-violet lights to underwater research', *Nature*, CLXXXV 259.

ZENKOVICH, V. P. (1956) 'Étude de la dynamique du littoral', *Proc. 18th Internatl. Geogr. Congress*, pp. 104–16.

—— (1958) 'Emploi des luminophores pour l'étude du mouvement des alluvions sablonneuses', *Bulletin d'information du Comité Central d'Océanographie et d'Étude des Côtes*, X 248–53.

The oven-cured epoxy resin used in the tests reported herein was Araldite 965-E, manufactured by C.I.B.A. Ltd, Duxford, Cambs., England. The catalysed polyurethane finishes were in the 708 range, manufactured by International Paints Ltd, London, England.

4 A Theory of the Development of Accumulation Forms in the Coastal Zone

V. P. ZENKOVICH

COASTAL accumulation forms are those positive relief forms which are determined by the action of certain waves and currents and are stable at a given sea-level and under given physico-geographic conditions. These forms can reach a length of several hundreds of kilometres and represent the concentration of several hundreds, or even several thousands of millions, of cubic metres of sediments. They have very varied shapes and outlines, and their formation often causes a modification of the general pattern of the coast.

The physico-geographic conditions noted above include, first of all, the hydrodynamic regime and the conditions of sediment supply. We assume that relatively coarse sediments (diameter more than 0·05 m.) are suspended in the waves; at certain stages they are dragged towards the sea bed by eddies, and the oscillations of the swell bring about their deposition in less than half the period of a wave.

The fundamental bases for a theory of the origin and development of accumulation forms were established by the author in 1945–6 (Zenkovich, 1945, 1946a,b,c). Research undertaken subsequently in the U.S.S.R. has confirmed them, although the incompleteness of these theories has also been indicated. A new series of factors in the study of accumulation forms has thus had to be taken into account (Zenkovich, 1950a,b, 1952a,b, 1957a,b). In this article these theories will be presented concisely.

The material involved in the development of accumulation forms can consist of:

(a) the products of abrasion from adjacent sectors of the initial coast;

(b) sandy alluvium or larger material supplied by streams flowing into the sea; and

(c) the products of sea-bed erosion mixed with the solid remains of marine organisms, and sometimes those of chemical origin (e.g. oolites). Wind-blown (aeolian) sediments may have some importance locally.

The size of the material may vary from fine sand (of less than 0·05 mm.) to blocks and large masses of rock. The larger material (gravel, pebbles, etc.) can only be transported by fluctuations in water movement. The finer the material, the more its movement is subject to storm currents, the reverse sea-bed current, the wave current and longshore drift.

Accumulation forms can also originate from other primary masses of loose material in the concavities of the initial coast. In this case a transgressing sea takes up and reworks the initial material and moulds the outer edge of the accumulation in accordance with topographic and hydrodynamic conditions. More often, however, accumulation forms result directly from marine action as a result of the longshore displacement of sediments by waves and currents, or because of the movement of sea-bed sediments into shallower waters or even directly on to the beach.

The lateral transfer of sediments can occur directly on the coast, in the breaker zone and also at varying depths on the sea bed. Accumulation forms appear where the movement of sediments is slowed down or stopped completely.

Onshore as well as lateral movement plays a part in their formation. The width of the submerged coastal slope, where these processes occur, depends on the wave parameters and on the size of the sediments. On sandy coasts it may reach several kilometres. The onshore displacement is greatest during the formation, or subsequent rearrangement, of the equilibrium profile. Generally speaking, onshore and lateral movements go on at the same time.

Morphologically, accumulation forms may be classified in two main ways: those that are entirely submerged, and those which rise above water. In this article the latter category especially will be discussed; it may be divided into five morphological groups (Fig. 4.1):

(a) *Attached forms*, linked directly to the initial coast (beaches, various emerged terraces, the infillings of coastal concavities and re-entrants).

(b) *Free forms*, in front of the coast, surrounded by the waters of a bay or the open sea, and extending outwards from the coast for a distance greater than the length of the part attached to the coast (spits, arrows).

(c) *Fringing forms*, a band of sediment extending along the coast

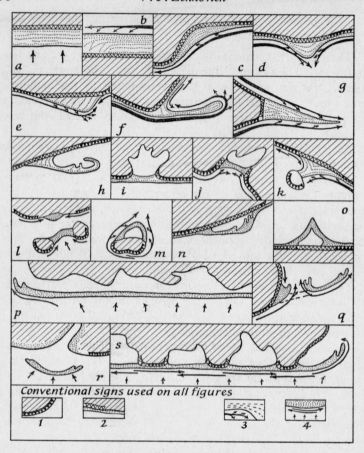

Fig. 4.1   The classes and main types of accumulation forms

Attached forms: *a* Emerged terrace formed by sediments from open water.
    b Same formation, supplied laterally.
    c Terrace formed by the infilling of a concavity (supplied laterally).
    d Symmetrical projection caused by supply from two directions.
    e Asymmetrical accumulation.
Free forms: *f* Spit (unilateral supply).
    g Arrow (bilateral supply).
    h Spit along a regular coast.
Connecting forms: *i* Beach at the mouth of a bay.
    j Beach inside a bay.
    k Asymmetrical tombolo.
    l Tombolo between two islands.

and linked to it at both ends like a lagoon bar (double fringing spits, cuspate forelands, etc.).

(*d*) *Hooked forms*, blocking bays in the middle or at their entrance, linking an island to the land, or two islands together (isthmuses, coastal barriers and tombolos).

(*e*) *Detached forms*, no longer linked to the initial coast (some coastal barriers and islets of accumulation).

From the *dynamic* point of view, accumulation forms can be either *stable* (maintained without new material reaching them) or *mobile* structures (the author once proposed the terms first-order forms and second-order forms respectively for these structures). The existence, situation and orientation of mobile forms are determined by the supply of materials. Should the supply cease or be reduced, the forms are eroded. The most varied mobile forms occur in places where longitudinal currents lose their capacity or bring in material less rapidly.

Sediment streams are characteristic of fairly regular coasts, along which material is displaced laterally. This process is the result of repeated migrations of material in opposing directions but with a net movement in one direction over a period of a year or more. A current has a given capacity and a given power. The first of these depends on the parameters, period and direction of the waves, and indicates the potential for transporting a given quantity of material

---

Fringing forms: *m* Looped spit behind an island.
                n Same formation on a regular coast.
                o Double spit.
Detached forms: *p* Coastal barrier beach.
                q Detached spit (relict form).
                r Island formed by the accumulation of sediments in front of a river mouth.
                s Barrier beach joined to headlands of the initial coast.

Conventional signs *used on all figures*:

1. Initial coast, limited by an active cliff.
2. The same, with a dead cliff on the left, and on the right a cliff fringed by a beach.
3. Accumulation forms, with coastal spits. The arrow along the coast represents the sediment stream. The thickness indicates its relative power. The short arrows pointing inland indicate the proportion of materials enlarging the fringing beach. The arrows pointing seawards indicate the loss of material towards deeper waters.
4. Accumulation forms, without coastal spits. The zone delimited by the two longitudinal arrows shows the area where sediments migrate in two directions. The short arrows pointing inland indicate the supply coming from the open sea.

ICD D

along a given coastal sector. This potential cannot be wholly achieved if there is little sediment near the coast. The true power of the current will be weaker than its capacity, and we have described such a current as unsaturated. To analyse the dynamics of the coast we shall also use the value of the resultant of the wave regime, that is, a vector of given magnitude and direction representing the sum of the energy of all the fluctuations in movement over a year. The projection of this vector on to the coastline determines the relative value of the capacity of the sediment stream at any given point (Zhdanov, 1951; Popov, 1957).

The sources of sediment that are able to form stable structures are most often situated in coastal concavities (terraces and isthmuses in bays, or tombolos between islands). The orientation of the resultant of the wave regime is rarely the same along the whole length of the formation because of wave refraction near headlands, etc. Hence the outer edge of such forms is often a regular curve and the resultant is normal at any point on the curve (Fig. 4.2, diagram 2).

The varying stability of the forms described is also an index of the degree of development of the equilibrium profile in front of them. Lateral transport is prevented by the flanking headlands, and sediments can only migrate to and fro along such a formation, the distances depending upon the degree of exposure and local topographic conditions.

The relation noted above between the orientation of the outer edge of certain accumulation forms and the normal to the direction of the dominant waves has been established by Cholniky (1927), Lewis (1934) and Schou (1959).

It follows from what has been stated that not all accumulation forms become stable. For instance, those that appear on a regular coast when the capacity of the sedimentation current decreases cannot be stable. If the supply decreases or ceases altogether, the waves reaching the coast obliquely will erode the existing terraces, accumulations or spits. For the same reason the free forms or fringing forms (those nearest open water) can only survive in rare instances when the supply ceases. Such are the remnant forms on an initial coast seen near an obstruction, or the accumulations and spits along the side of an island sheltered from the waves (Fig. 4.2, diagram 2).

From the genetic point of view, coastal accumulations may appear first where the supplying currents slow down (see below for the four possible causes of this phenomenon) or second, when sediments are thrown up from the sea bed. These two processes may act together to produce forms of mixed origin. The stable forms mentioned above which form in protected areas of the coast during a marine transgression can be considered as one of the groups of relict forms under present conditions.

It must be emphasised that similar or even identical forms can differ both in their mode of formation and in the origin of their materials. Thus, for example, emerged terraces may be formed from materials derived from the sea bed (Fig. 4.1(*a*)) or from the sediments carried by a lateral current (Fig. 4.1(*b*),(*c*)). Similar examples will be discussed later.

The displacement of sediments towards the coast leads to the appearance of emerged accumulation forms (coastal bars, attached terraces) or submerged forms (several types of spits which are sometimes very extensive) (Zenkovich, 1957*b*). When the ends of these accumulations are attached to capes, or even if attachment occurs at one or more intermediate sections, then morphologically they become isthmuses. These features, fed directly by bottom sediments, only preserve their characteristic form if the resultant of the wave direction stays normal to the coast. But the material from the sea bed is sometimes picked up by lateral movements and thus becomes part of a sediment stream. Currents may also exist on the outer edge of barrier beaches; this leads to the growth of secondary accumulations at their extremities.

When sea-level remains stable, the most gentle of submarine slopes can only furnish a limited quantity of sediment once the equilibrium profile is established. In such conditions neither barrier beaches nor terraces can develop to any width. Different conditions occur when a flat terrestrial surface of unconsolidated deposits is invaded by the sea. The material accumulating on the upper part of the submarine slope also migrates coastwards and increases continually in volume at the expense of the terrestrial deposits, which sooner or later are eroded by the waves. Thus barrier beaches and submarine bars several kilometres long can develop, in the construction of which lateral displacement has not taken part (Zenkovich, 1957*b*).

A decrease in the capacity of sediment streams which build accumulation forms can take place under the following conditions:

(*a*) A curve or re-entrant in the direction of the initial coastline (re-entrants or concavities are infilled).

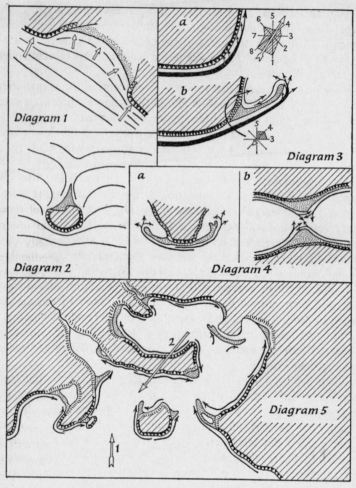

Fig. 4.2

1. Stable accumulation form (emerged terrace) in a bay-head

The lines drawn on the sea are wave fronts. The double arrows represent the wave directions.

2. Arrow formed behind an island in the zone sheltered from the waves

This form is stable when the wave crests reach it at right-angles.

3. A secondary sediment stream linked to the growth of the spit

(*a*) Initial coast and a sediment stream which loses its power as it rounds the corner. On the right are the wave rose and the wave resultant.

(*b*) Sediment streams after the formation of the spit. In the zone sheltered from (*b*) A landward curve in the coastline (sediments pass around a headland and form accumulations or arrows).

(*c*) A decrease in the wave energy which generally occurs where the water surface decreases (in a bay or strait). In this case asymmetrical accumulations appear which can later become arrows or fringing forms.

(*d*) A section of the coast interrupted by an island, cape or bank (accumulations will form and increase in size to give arrows, isthmuses and tombolos).

According to the contour of the coast, either one sediment stream or two opposing streams may be slowed down in the same sector. In the first case accumulation forms supplied unilaterally will appear (simple forms), which are generally oblique to the main direction of the coast (Fig. 4.1(*e*),(*f*),(*h*),(*k*),(*n*)). In the second case double forms, supplied bilaterally, will grow which are sometimes perfectly symmetrical (e.g. Fig. 4.1(*d*),(*g*),(*m*),(*o*)).

There is no clearly defined limit between simple and double forms, because on the inside of the deposits and on the sector belonging to the initial coast there is almost always a secondary sediment stream of little power which runs in the opposite direction to the main stream. This follows because the growing accumulation form protects the coast behind it from the dominant waves. However, the subordinate waves still continue to act and thus modify the value and direction of the resultant (Fig. 4.2, diagram 3). Indeed simple forms can scarcely be distinguished from double forms unless particular wave systems predominate.

If the sediment streams diverge at a small projecting headland of the initial coast, two paired deposits can form on its flanks as a result of the erosion of the cape or island (Fig. 4.2, diagram 4(*a*)). Paired forms of different origins are frequent in straits and narrow

---

the waves, behind the spit, only waves of directions 3, 4 and 5 remain active. Their wave rose and wave resultant have been plotted.

4. Paired accumulation forms

(*a*) Spits established on a headland approached by two opposing wave systems.
(*b*) Projections formed in a strait where waves act on both sides.

5. Schematic diagram of sediment streams and accumulation forms on a dissected coast

Arrow 1. The resultant of the open-water motion.
Arrow 2. The resultant of the winds locally pushing waves towards the interiors of the bays.

bays where a projecting deposit formed on one side leads to the slowing down of the currents or to the migration of sediment to the other side (Fig. 4.2, diagram 4(b)). In this case one projection may be considered as primary and the other as induced.

It is along irregular coasts that converging and diverging currents are commonly found. The coastal irregularities face in many directions and are therefore attacked by waves which vary both in length and in their direction of approach. The currents are also split up, so that each part of the coast is followed by what may be regarded as its own local current Fig. 4.2, diagram 5, shows the sediment streams actually observed on a dissected coast.

When the initial coast changes direction, the appearance of accumulation forms depends on the orientation in relation to the wave regime of the sector being eroded, along which a sediment stream moves, and also on the degree to which this current is saturated. As these relationships have often been described in the literature (Zenkovich, 1945, 1946) the discussion will be limited to Fig. 4.3, diagrams 1 and 2.

In order to understand these figures, a value of the angle α between the coast and wave direction (or of the wave resultant) must be considered, which corresponds to the fastest lateral transfer of material (or the capacity of the sediment stream). This angle (called ϕ by Soviet authors) varies between 40° and 50°, but can be taken as 45° on average. De Lamblardie had already tried to find this angle theoretically in 1789. Experiments have been carried out in wave tanks to determine this angle (Sauvage de Saint-Marc and Vincent, 1955). The problem was also considered from the aspect of energy balance by Longinov (1958).

Imagine a sector of an initial coast AB along which flows a sediment stream. At point B the coast changes direction at a salient or re-entrant. Fig. 4.3, diagram 1, shows all the possible new directions that rays from this point can take. If AB, along which a saturated sediment stream flows, is placed at less than angle ϕ in relation to the wave resultant (Fig. 4. 3, diagram 1 III), the speed decreases, whatever the direction of the change in the coast may be. A seaward deviation gives a re-entrant which fills with sediment. In relation to the reduced speed deposition continues, theoretically for each phase of the growth of the formation, to the right along the initial coast as far back as the source of the material. The direction of the outer edge of an accumulation coast along which deposits are forming can be found by establishing the relationship between the speeds of dis-

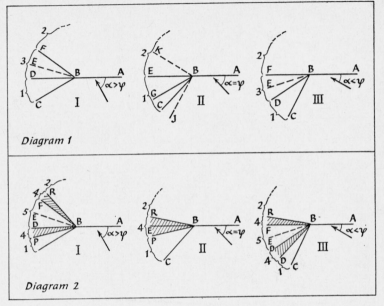

Fig. 4.3

1. *Diagram of the appearance of accumulation forms when the orientation of the initial coast is modified*

    I. Sector where infilling takes place in a coastal re-entrant.
    II. Sector where a coastal projection is rounded.
    III. Sector of erosion.

    2. *The same, with an insufficient supply of sediment*

    I. Sector where supply is sufficient.
    II. Sector where erosion is increased. Other indications as for Diagram 1.

placement along the new sectors (*BC*) and the old sectors (*AB*). If the second side of the re-entrant angle is formed by *BC*, along which there is no displacement of sediment, the outer edge of the deposit grows parallel to *AB*. This implies a general advance of the sector *AB*, mainly to the right of it. If the angle is smaller, a new system establishes an equilibrium, and in so far as the material extends beyond point *C*, the coast maintains a certain obliquity (Fig. 4.1(*c*)). Pellenard-Considère (1956) attempted a theoretical solution of the problem by applying it to the infilling of the angle in front of a harbour wall. The sector *EBC* may be called the filled sector. The ray *CB* and the direction of approach of the waves are at right-angles. If the angle increases still more (e.g. *JB*), a sediment current from an opposed direction flows along the ray and accumulation will take place from two distinct sources.

When a coast turns in a landward direction, a cape is formed. As long as the displacement of sediments is weaker in this direction (e.g. *BK*) than along *AB*, sediments accumulate immediately behind point *B*. The constant supply of material forces the accumulation away from the coast in the form of a spit extending in the direction of the movement of a saturated sediment stream of a given power. It forms a prolongation of the initial coast along *BE*. The spit cannot grow either to the left or to the right of this direction. In the first case the material would be eroded more rapidly than it was supplied. In the second case the rate of supply would be greater than that of erosion. The accumulation form will only be stable in a position where each part of the spit loses exactly as much material as it receives. Thus, since in each case the free forms can only be in equilibrium if they run in the same direction as the saturated current of a given power, their spatial orientation is logical. Sector *EBK* situated above the line *EB* has been called the sector of headland-rounding. When the coast and the resultant are at an angle to one another which is either greater or smaller than ϕ, the conditions of accumulation growth vary noticeably. In the first case (Fig. 4.3, diagram 1 III), $\alpha < \phi$. If the coast curves seawards, the capacity of the current first increases, reaches its maximum for angle ϕ (line *EB*) and then decreases to its original value (line *DB*). In the sector *DBF* (the erosion sector) the coast will be eroded.

The infilling of the angle only starts when there is a more marked curve (near the base of *DB*). In these conditions the process will differ from that shown in Fig. 4.3, diagram 2 II, because the limit to which the sediments accumulate will immediately take the direction ϕ (i.e. parallel to *EB*); later the primitive triangular form will increase and maintain a similar shape at each new stage, and will extend only a limited way on the sector *AB*.

This ordered growth of accumulation forms can be observed on a small scale on a regular coast when waves approach it at a very acute angle. Small spits are then formed at about 45° to the wave direction. Johnson (1919) placed these forms in his offset group. On a large scale, the process ends in the development of formations characteristic of the Sea of Azov (see below).

Similar conditions to those already examined occur in the sector *EBA*, because a spit can only grow in the direction of a saturated current of given power (*BF*) along the extension of line *AB*.

When the initial coast forms an angle greater than ϕ (Fig. 4.3, diagram 2 1), the erosion sector appears on the landward curve in the coast. In such a case a spit will always be turned landwards at an angle (*DBF*) which is greater if the difference between α and ϕ itself increases.

If an unsaturated current flows along the initial coast, the decrease in capacity in a sector of given width will not be accompanied by sedimentation. On the three diagrams (Fig. 4.3, diagram 2) deposition will only start on the outside of rays *PB* and *RB* where the current becomes completely saturated. On the other hand, if both the rate of displacement and the capacity increase, the erosion accompanying the movement of the unsaturated current along the initial coast *AB* will also increase. Thus sectors of saturation (*PBR*, *PBD*, *EBR*) should be shown on the diagrams, and the erosion sector (*DBF*) should here be termed the sector of increased erosion. The growth of accumulation forms only starts at the outer edges of the sectors of saturation.

All these rules are best illustrated when accumulation forms are built of large-sized materials such as pebbles. Sandy accumulations are influenced by the transporting action of the current, which tries to preserve the direction of the initial coast.

The differences between stable and mobile forms may be emphasised by further consideration of Fig. 4.3, diagram 1. Assume that the supply of material ceased when the infilling of the angle reached point *C*. Some material then continues to migrate beyond this point, and the volume of the deposit will diminish (Fig. 4.4, diagram 1). The outline of the remaining material will not, however, become rectilinear. It will curve because the initial segments will accommodate themselves to wave refraction, or it will present large scale concavities which will limit the amount of sediment movement in relation to the annual cycle of waves having different directions and forces.

If, after the formation of a spit from point *B*, sediment is no longer supplied to it, the material will continue to move along the spit. The proximal part of the spit will first feed the distal end, but later the sector of maximum loss will be broken through and the spit will become an island. Its ultimate destiny will depend on the depth of the sea. The island will acquire recurved ends (secondary spits) and little by little will be displaced in the direction of the wave resultant, until finally it disappears, by which time all the material will have been used to build up a submarine bank (Fig. 4.4, diagram 2).

ICD D 2

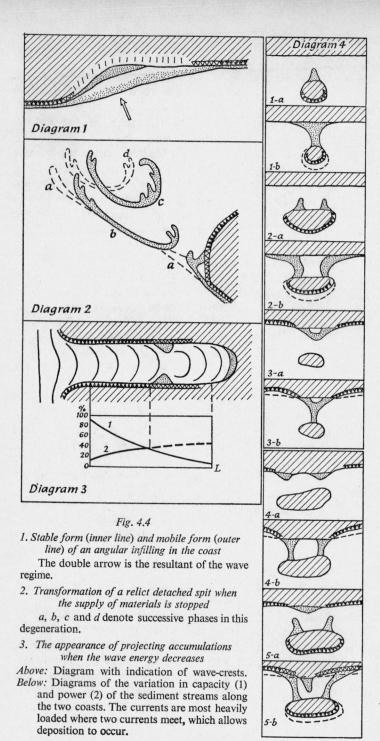

Fig. 4.4

1. *Stable form (inner line) and mobile form (outer line) of an angular infilling in the coast*

   The double arrow is the resultant of the wave regime.

2. *Transformation of a relict detached spit when the supply of materials is stopped*

   a, b, c and *d* denote successive phases in this degeneration.

3. *The appearance of projecting accumulations when the wave energy decreases*

   Above: Diagram with indication of wave-crests.
   Below: Diagrams of the variation in capacity (1) and power (2) of the sediment streams along the two coasts. The currents are most heavily loaded where two currents meet, which allows deposition to occur.

The development of accumulation forms occurs much more simply in other conditions (noted on p. 99). Obstruction implies the total or partial protection from wave action of any sector of the initial coast, along which sediments migrate or where a sediment stream exists. The obstruction may be an island, a submarine bank or a prominent headland. Protected by these features, materials accumulate and form a salient – a spit or an arrow. The size and degree of asymmetry of these forms depends on the amount of material available and on the structure and orientation of the wave resultant. Under favourable conditions the accumulation form can attach itself to the obstruction. This process will be demonstrated in greater detail below, in the analysis of tombolo formation.

The effect of a reduction in wave energy can best be seen in long narrow bays with parallel sides (Fig. 4.4, diagram 3), on which, at given points, asymmetrical accumulations grow up without the intervention of any external factor (for example, an obstruction or any change in coastal direction). The localisation of these forms depends on the degree of saturation of the sediment stream along the bay sides and also on the shape of the bay opening. The primary location of the accumulation is generally situated in a zone where the current is completely saturated. The power of the current nearly always increases towards the bay-head, as a result of the erosion of the long sides, although the current capacity decreases because the wave parameters decrease. If the variations of these factors are represented by curves (Fig. 4.4, diagram 3), their intersection indicates where the primary accumulation takes place. The subsequent evolution of the deposit sometimes involves its displacement towards the bayhead.

In the area protected from the waves (the shadow zone) between an island and the mainland, accumulation forms grow up in different ways according to the origin of their component materials. On a mainland coast, even where there is a current or where noteworthy migrations of sediment take place, accumulations can occur only where the island gives a shadow zone of more than 15° (Knaps, 1950). Then a form of obstruction appears (Fig. 4.4, diagram 4). If the island undergoes intense erosion, and if its size is reduced, a

---

4. *The initial (a) and final phase (b) in the formation of tombolos of different types and origins*

1 and 2 come from the products of the abrasion of the island, 3 and 4 from those of the mainland, 5 is polygenetic.

narrow accumulation, fed from two sides, will form behind the shadow zone. An accumulation behind the island, and the growth of a projection from the mainland, can eventually block a strait and form a tombolo. The two processes can occur simultaneously. Behind larger islands two spits may form, which may reach the mainland coast. The definitive aspect of any tombolo, its degree of asymmetry and the origin of the materials, will be determined by local conditions. Triple tombolos may form on a coast or an island if, for a time, accumulation takes place at three points simultaneously. Perhaps one of the group will develop more rapidly than the others. The middle one of a triple tombolo, however, is always a relict form.

When an island is completely isolated by erosion, a detached remnant may remain behind it (Nichols, 1948), and can be pushed towards the coast and become moulded to it. The submarine base of the island plays an important obstructive role, and thus some of the sediments, which at first formed a prominent tombolo, are preserved in a much reduced form.

The processes which give rise to fringing forms are most diverse. In the simplest case two spits join up behind an island and enclose a lagoon. This has been called a looped spit (Fig. 4.1(m)). Asymmetrical fringing spits generally appear in bays in places where the wave energy decreases and also when the sediment stream becomes impoverished. The advance of the accumulation into the bay thus slows down and local waves, normal to the shore, predominate over others. This leads to the growth of the end of the spit towards the coast and ceases when the two are joined. The appearance of fringing forms is sometimes characterised by a reduction in the size of the materials, which in turn modifies the ways in which they are set in motion by waves of different force. Such a mechanism was described by the author (Zenkovich, 1950*a*) from his studies of a bay in eastern Kamchatka. A fringing form can grow from a pre-existing massive form after a rise in sea-level (see below).

Marked differences between the initial stages and the subsequent development of accumulation forms can be explained by current action. On sandy coasts the main mass of sediment is displaced parallel to the coast along the sea bed in the breaker zone, whilst on a pebbly coast the displacement takes place in a narrow zone and along the beach. This is why sandy spits are widened seawards; the growth of their emerged parts occurs on the distal part of the spit (Bülow, 1954).

Where the coast curves inland, the sediments are sorted in

accordance with their hydraulic dimensions by current action, and a wide dispersion fan is formed (Zenkovich, 1957*a*). The finest particles continue to be carried by the current, but the largest are deposited to form a beach.

When a narrow re-entrant is filled with fine sand, a submarine spit is formed in the breaker zone and becomes attached to the adjacent coast; subsequent accumulation of new sediments leads to its emergence.

Whatever their mode of origin, forms pass from one morphological type to another during their growth and development. For example, a simple accumulation becomes a spit, which, on reaching the opposite side of a bay or on meeting a similar paired form, becomes an isthmus (Fig. 4.5, diagram 1). If, as a result of special hydrodynamic conditions, a spit becomes reattached to the coast from which it springs, then it becomes a fringing form, and so on. There is thus an increase in the size of the form and in the volume of the constituent materials.

However, the general development of a coast is usually a simplification in outline and a decrease in the power of the sediment streams, which can lead to a reduction in the size of accumulations so that they may change into different morphological types. Thus an isthmus may become a detached form or be broken up into two spits, which in turn degenerate into amorphous masses.

Mobile forms may disappear completely (Boldyrev, 1958; Zenkovich, 1953). Their constituent materials are dispersed over the sea bed to accumulate again in some other shape at a more stable point along the coast.

Certain types of accumulation forms can only originate under the action of two or even three wave systems from different directions. This process is especially clear in the development of accumulations along a regular coast, where wave energy is reduced. Where all the material cannot be carried by the stream, sediment is deposited in a narrow terrace which may become a salient.

The dominant waves displace material along the accumulation so that its end grows. Waves that are more or less normal to the coast move part of the material along each side towards the coast (Fig. 4.5, diagram 2). Material protected from the waves remains *in situ*. The waves from an opposite direction can also add small quantities of material from the bay itself or from the opposite side of a strait. Thus, as the form increases in size, it is displaced towards the interior of the bay or away from the mouth of a strait.

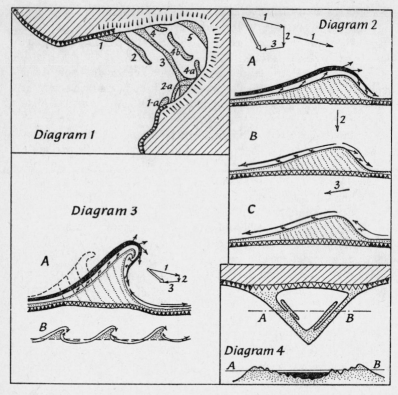

Fig. 4.5

1. Development and degeneration of a barrier formed in a bay

Two paired projections formed by the accumulation of material (1 and 1*a*) become spits (2 and 2*a*), then an isthmus (3) established across the bayhead. If the supply of material diminishes, the isthmus breaks up into a number of units (4, 4*a* and 4*b*), then all the material is pushed up to the bayhead.

2. The dynamics of a projecting asymmetrical accumulation formed along the coast of a bay

Above left: The polygon of the chief directions of the swell and their resultants.

(*a*) The action of the swell. A powerful sediment stream from the bay mouth is enriched by material eroded from the left side of the accumulation and deposits material on its right side.

(*b*) The action of the waves erodes the end of the accumulation and, from both sides, the material migrates towards the initial coast.

(*c*) The action of weak waves coming from the bay itself.

3. The dynamics of an Azov-type spit

(*a*) A single spit with a system of coastal banks and a supplying current; on the

The general shape and degree of asymmetry of an accumulation form depends on the relative intensity of the wave systems already noted. If normal waves are rare or weak, the distal end of the accumulation is pointed; if waves have sufficient power, the end is rounded. With opposing wave systems of equal intensity, as in the middle of a strait, the accumulation form becomes symmetrical and is not then displaced.

Under certain wind conditions, spits or arrows may form on coasts facing the open sea, and occupy a position corresponding to line *BE* on Fig. 4.3, diagram 1ɪɪɪ. Since such formations are well developed on the northern coast of the Sea of Azov, they have been given the name Azov-type spits in Soviet literature. It is important to note the following characteristics (Fig. 4.5, diagram 3). The development of Azov spits is a rhythmic process; they form in series, equidistant from each other (for example the lagoons of the Chukchi Sea – (Boudanov, 1956; Zenkovich, 1952*b*), or they may be separated by distances which vary regularly with the increase in the power of the waves. Near a coast where such spits are formed, the wave resultant can be close to the normal, but consists of two chief vectors which are parallel to the coast or intersect it at a very small angle. Azov-type spits are always displaced in the direction of the sediment stream.

The erosion of salients on the initial coast affects the accumulation of associated forms. The forms migrate, their orientation is modified and several may join together. Individual characteristics are thus lost and their outlines become simpler. If, for example, the sea cuts away the headlands enclosing bays, accumulations may form a single spit which will continue to migrate at the same rate as the coastline. Such spits are generally narrow (Fig. 4.1(*i*)).

When rapid erosion of headlands in unconsolidated deposits helps to supply accumulation forms, a special phenomenon may be observed. What may be called abrasion and accumulation couples are formed. Any modification of their outlines is simultaneous in both time and space. These couples undergo rotation, and in so

---

right side is the polygon of wave directions, on the left is the previous position of the spit which is displaced towards the right.

(*b*) A coastal sector with a group of such spits.

4. A rise in sea-level changes a coastal accumulation into a double fringing spit

Section *AB* shows in black the muds deposited in the resulting lagoon.

doing tend to approach the normal or the wave resultant. During this process the fulcrum point, where the coast provisionally remains stable, migrates in general from the zone of erosion towards that of accumulation. This phenomenon has already been described by Davis (1912) and Johnson (1919).

The complication of the outlines of accumulation forms, and their transformation into fringing forms, for example, may be caused by a relative lowering of the coast, whilst at the same time accumulations continue to grow. The oldest ridges in the interior of the form come increasingly close to the new sea-level, until they are completely submerged (Fig. 4.5, diagram 4). This mechanism is best seen on projecting deposits and on wide spits and barrier beaches. When the latter increase simultaneously on both the seaward and lagoon sides, they become double, that is, composed of two parallel bands of ridges (Fig. 4.6, diagram 1(*a*)). As it is lowered, a coast bordered by an accumulation terrace can become a lagoon coast if the back of the terrace is submerged and if the front continues to increase in height (Fig. 4.6, diagram 1(*b*)). Such coasts have been described by the author (Zenkovich, 1954). The forms shown in Fig. 4.6, diagram 1, have been described both by the author (Zenkovich, 1952) and by Fisher (1955).

As accumulation forms grow upwards, secondary forms may appear on them or be caused to form on adjacent coastal sectors (Fig. 4.6, diagram 2). Examples of the first case include the infilling of re-entrant angles at the inside end of a spit or isthmus, and also the series of projections on their inner sides near their distal ends. In all similar cases the mechanism already described can occur on a smaller scale. Induced forms may grow on the original coast as a result of the obstruction caused by the growth of a major feature across the mouth of a bay. Paired forms of similar origin may be built on both sides of a strait.

Secondary forms can also appear in the waters enclosed by a fringing structure or in lagoons that originate in impervious accumulations as a result of a lowering of the coast. Some secondary forms are relics (Fig. 4.6, diagram 3), as, for example, the recurves in barrier beaches or those at the ends of spits that existed in an early stage in the development of the major feature. They may last for a long time if the waves in the basin are weak.

The shapes of accumulations (the character of the curves, the presence of secondary or relict forms, etc.) are determined by factors which complicate their evolution. Here may be noted:

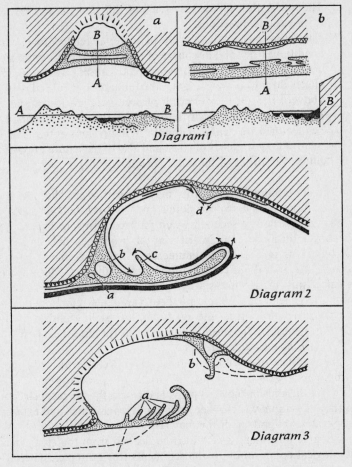

Fig. 4.6

1(a). Double spit developed from a simple spit as a result of a rise in sea-level

On the section *AB* marine deposits are shown in large dots, with deposits derived from the bay in finer dots.

(b) *The coast bordered by a terrace is transformed into a lagoon coast as sea-level rises*

Alluvial deposits and lagoon muds are shown in black.

2. Secondary and induced accumulation forms caused by the growth of a spit

3. Relict accumulation forms

(a) Recurved ends.

(b) Fringing spit which is no longer being supplied with material. The dotted line shows its original outline.

(*a*) The alternation of wave systems of different parameters and from various directions.

(*b*) The variations in the supply of material related to climatic conditions (the solid discharge of rivers, the speed of abrasion, etc.) or to other physio-geographic causes. These indirect effects do not, however, upset the fundamental rules described above. In short, the structure of accumulation forms and the relative ease with which the origin of their constituents may be identified are clear and important indications of the development of long stretches of coast and also of the nature of possible future variations of the coast (Leontev, 1954).

Complex accumulation forms may appear where the effects of two or more of the factors noted on p. 99 are felt, or after the growth and fusion of secondary and induced forms, or again where migrating forms join up when coastal headlands are subject to erosion. It is impossible to enumerate all the possible variations, because the number of combinations of only three or four elements is sufficiently high. This is why, when complex forms are studied from the genetic or dynamic point of view, their component parts must be separated and the evolution of each established. The methods described above allow this analysis for almost every feature which may be encountered.

The outlines of accumulation forms, the angle at which their component ridges run relative to one another and the presence of hooks at the ends of the spits provide the data for the reconstruction of their development. Hooked spits imply open water and extensive longitudinal migration of the material.

Valuable information may also be derived from the study of the lithological composition of the sediments (for example, material of organic origin, minerals, the presence of rock debris emanating from a particular part of the coast, or fluvial alluvium). It is also important to consider the volume of sediment accumulated in relation to the length of the coast subject to erosion; this may suggest whether the supply is derived primarily from the sea bed or from the coast.

In retracing the development of accumulation forms, it must be remembered that their initiation and early stages may have taken place under very different topographic and hydrodynamic conditions. The trend of old coastal barriers almost always reveals the extent to which the initial coast has changed its orientation. The directions of the former currents can be established when the original position of the initial coast has been noted and when the present forms are ignored. Corresponding corrections can then be made for the values of wave power and the overall wave resultant even if only a few variations and elements of former conditions are found. The variations involved can be quite marked where important accumulation forms have developed. Hence it would be wrong to analyse the conditions pertaining to the initiation and first stages of such forms if we made use of the present wave and current systems.

Some specialists also err when they try to establish the relationship between the dominant wave direction and the direction of the whole, or of the outer edges of all accumulation forms. Theory shows that the angle ϕ (corresponding to the maximum speed of sediment migration) is the essential element in this problem. The degree of saturation of the sediment stream and the orientation of the initial coast may make the directions in which the accumulation forms grow extremely varied, and may be modified as the coastline evolves. It must also be remembered that the orientation and outlines of accumulation forms are governed by the intensity of the alternating action of a spectrum of waves varying in power and direction (normal to the coast, longitudinally, or in both directions).

The greatest number of accumulation forms, with the greatest variety of outlines, occur on indented and dissected coasts that are subject to relatively rapid erosion. The localisation and even the initiation of these forms are the results of a mechanism which first established local centres of supply and then spreads a mantle of sediment over the submerged parts of a whole coastal sector. The forms that rise above water are then built where the accumulation, at the time, is most intense.

The material supplied by the simplification of an indented coast combines with that released by the disappearance of a series of accumulation forms; the whole can then migrate to neighbouring sectors.

REFERENCES

BOLDYREV, V. L. (1958) 'Erosion of littoral forms of accumulation, e.g. the Kertch Channel', *Work of the Inst. of Oceanography, Acad. Sci. U.S.S.R.* (Trudy 10 AN), XXVIII.

BOUDANOV, V. I. (1956) 'The formation and development of spits of the Azovian type', *Work of the Commission of Oceanography, Acad. Sci. U.S.S.R.*

BÜLOW, K. (1954) *Coastal dynamics and formations in the Southern Baltic Sea* (Berlin).

CHOLNIKY, E. (1927) 'Early transformations of the coastal areas', *Peterm. Geogr. Mitt.*, pp. 7–8.

DAVIS, W. M. (1912) *Die beschreibende Erklärung der Landformen* (Berlin-Leipzig).

FISHER, R. L. (1955) 'Cuspate spits: St Lawrence Island', *J. Geol.*, LXIII, no. 2.

JOHNSON, D. W. (1919) *Shore Processes and Shoreline Development* (New York).

KNAPS, R. Y. (1950) 'Effect and application of breakwaters as protective installations on sandy coasts', *Bull. Acad. Sci. Latvian S.S.R.* (Riga), no. 7 (36).

DE LAMBLARDIE (1789) *Report on the Coast of Upper Normandy, etc.* (Le Havre).

LEONTEV, O. K. (1954) 'Morphological analysis: one of the basic methods for studying coast formation', *Public records of the Univ. of Moscow, Geogr. Series*, no. 10.

LEWIS, W. V. (1934) ' The effect of wave incidence on the configuration of a shingle beach ", *Geogr. J.*, LXXVIII, no. 2.

LONGINOV, V. V. (1958) 'The dynamics of the breaker zone and of the littoral zone', *Work of the Inst. of Geol. and Geogr., Acad. Sci. Lithuanian S.S.R.* (Vilnius) VII.

NICHOLS, L. (1948) 'Flying bars', *Amer. J. Sci.*, CCXLVI 96–100.

PELLENARD-CONSIDÉRE, R. (1956) 'Essay on the theory of the evolution of sand and pebble beaches', *Compte Rendu IV de l'Hydrologie*, G III, Report No. 1.

POPOV, B. A. (1957) 'The determination of the results of movement of the waves', *Annals*, 10, XXI.

SAUVAGE DE–SAINT MARC, G., and VINCENT, G. (1955) 'Coastal movement: formation of spits and tombolos', *Coastal Engineering*.

SCHOU, A. (1959) 'The Danish moraine archipelago as a research field for coastal morphology and dynamics', *Proc. 2nd Coast. Geogr. Conf.* (Louisiana).

ZENKOVICH, V. P. (1945) 'Formation of accumulation structures modifying the direction of the initial coastline', *Proc. U.S.S.R.* (Trudy Inst. Okeanol. Akad. Nauk. S.S.S.R. 10) XLVIII, no. 5.

—— (1945–6) 'Theory of the formation of spits and other coastal structures', *Proc. Nat. Inst. Oceanography*, nos. 1–17.

—— (1946a) 'Formation of coastal accumulation structures with debris from the coast', *Dokl. Akad. Nauk. S.S.S.R.* LIV, no. 4.

—— (1946b) 'Recent studies on the dynamics of coasts', *Bull. Nat. Geogr. Soc.*, no. 5–6.

—— (1946c) *Dynamics and Morphology of the coasts* (Ed. Morskoy Transport, Moscow).

—— (1950a) 'Formation of bars in rings', *Dokl. Akad. Nauk. S.S.S.R.* LXXI, no. 3.

—— (1950b) 'One method of lagoon formation', *Dokl. Akad. Nauk. S.S.S.R.* LXXV, no. 4.

—— (1952a) 'Double and tombola coastal bars', *Piroda* (*Nature*), no. 2.

—— (1952b) 'Development of lagoons', *I.V.G.O.*, no. 5.

—— (1953) 'A type of coastal formation in the process of erosion', *I.V.G.O.*, no. 2.

—— (1954) 'Dynamic classification of sea coasts' (Trudy Inst. Okeanol. Akad. Nauk. S.S.S.R. 10) XVI.

—— (1957a) 'The sorting of sedimentary material on the edges of spits' (Trudy 10 AN) XXI.

—— (1957b) 'The origin of the coastal bars and lagoon coasts' (Trudy Inst. Okeanol. Akad. Nauk. S.S.S.R. 21) XXI.

ZHDANOV, A. M. (1951) 'Determination of the power of a sediment stream by direct observation', *Bull. Acad. Sci. U.S.S.R., Geophys. Series*, no. 2.

5 The Relationship between Wave Incidence, Wind Direction and Beach Changes at Marsden Bay, County Durham

CUCHLAINE A. M. KING

INTRODUCTION

THE observations on which this paper is based covered the period from January 1949 to December 1950. They included weekly measurements of the beach profile during part of the period at two selected points and were used to correlate beach changes with variations of wind force and direction, wave characteristics and other relevant data. The paper is presented in four parts :

(i) Characteristics and location of the beach.
(ii) Significant factors governing wave action in the locality and prevailing weather conditions.
(iii) Analysis of the observations, with reference to relevant wave-tank experiments.
(iv) Generally applicable results and conclusions.

CHARACTERISTICS AND LOCATION OF THE BEACH

Marsden Bay is a small sandy bay lying between the Tyne and the Wear on the north-east coast of England. The beach is backed by Magnesian Limestone cliffs throughout its length. The sea reaches to the foot of the cliffs at high spring tide: erosion of the cliff is therefore able to proceed slowly under marine action, and this helps to maintain the steep cliff profile.

The bay is about 1,200 yds long and is separated by several miles of cliffs and rocky foreshore (to north and south) from the nearest sand beaches in the locality, those at South Shields and Seaburn. It does not seem likely that sand reaches the bay by longshore drifting from either side, so that the sand on the beach can only be moved within the bay or to the offshore zone.

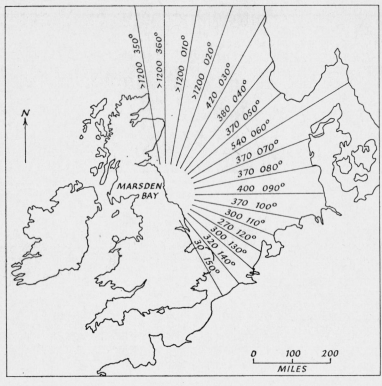

Fig. 5.1   The exposure of Marsden Bay, showing fetch in miles

The beach faces north-eastward and is exposed to the North Sea. The maximum fetch is in a northerly direction and extends to the Arctic Ocean, a distance of at least 1,200 miles. The length of the fetch in other directions is shown in Fig. 5.1. The prevailing wind, which is south-westerly, is an offshore wind; the dominant wind is, by contrast, the onshore or northerly wind. The beach is thus well situated to demonstrate the effect of the prevailing offshore wind as opposed to the dominant onshore wind.

SIGNIFICANT FACTORS GOVERNING WAVE ACTION

(a) The tides

The tides in this part of the North Sea have a range of about 14 ft at spring tide and 7 ft at neap tide. The tidal current flows to the south when the tide is near high water, and to the north near low water. Thus on the part of the beach exposed at low water the drift due to the tidal currents must be predominantly to the south, while the area which comes under the influence of the north-flowing currents at low water lies farther seaward. The tidal currents are not, however, sufficiently powerful to move much sand unless it is first thrown into suspension by wave action.

(b) Beach material

The beach is composed of a rather coarse-grained sand. The distribution of particle size is symmetrical about the median diameter, 0·37 mm. at the north end of the bay and 0·35 mm. at the south end. At the north end of the bay rocks are exposed near low water at some seasons. Some rather large shingle is at times exposed in the centre and south of the bay, but the amount of shingle varies with weather and wave conditions. Further complications are caused by the presence of stacks on the beach and by the fact that at the highest spring tides the waves reach to the cliffs at the back of the beach.

(c) Wind and wave data

In order to analyse the cause of the beach changes, wind and wave data are required. The wind observations were obtained from the daily weather report in which six-hourly observations for Tynemouth are recorded. This station is about four miles north of Marsden Bay and therefore gives an adequate record of the wind in the vicinity of the beach. To obtain a clear general picture of the wind between successive beach surveys, wind roses, showing the wind in one of eight directions and three speeds, were constructed for the period between two profiles (Fig. 5.6).

It is generally agreed that the dimensions of the waves are fundamental factors on which depend the amount of sand moved and the gradient of the beach. Observations of wave period were made by using a stop-watch. The wave height at break-point was estimated on all occasions when the beach was visited. From the wave period the deep-water wave length could be calculated, using the accepted formula $l = 5·12t^2$, where l is the deep-water wave length in feet and t the wave period in seconds (Beach Erosion Board, 1942). A wave is said to be in deep water when the water depth is more than half the wave length. The deep-water wave height was obtained from the breaker height by means of a set of curves, which took the wave length into account. Using these figures the wave steepness, or height-to-length ratio, could be found and the wave energy, given by the formula $e = 41h^2t^2$, could be calculated. e is the wave energy in foot-pounds per foot of wave crest per wave length, h is the deep-water wave height and t the wave period in seconds.

The average of all the wave periods recorded during the sixteen months of observations was just over 8 seconds. This figure, which corresponds to a deep-water length of 345 ft, indicated that the waves reaching the beach had travelled for a considerable distance to arrive as swell. The longest waves came from the north, the direction of maximum fetch. The low steepness value also pointed to the waves being far-travelled swells, and this was particularly noticeable during periods when the prevailing westerly winds were blowing offshore. The effect of a headwind on a wave system is to reduce the height of the waves without affecting their length, which leads to a small steepness value. Onshore winds, on the other hand, tend to increase wave heights and generate local seas which are normally shorter, resulting in an increase of wave steepness. Under these conditions two sets of waves are often present at the same time, the shorter seas combining with the longer far-travelled swell.

ANALYSIS OF THE OBSERVATIONS AND WAVE-TANK EXPERIMENTS

(a) The surveys

Two profiles were levelled at weekly intervals, one at the north end of the bay and the other at the south. The tide heights, calculated from the Admiralty tide tables from the observed water-level at a known time, were entered on the profile. The surveyed profiles enabled changes in the surface level of the sand to be measured accurately at any point on the profile, and beach gradients could be ascertained. Some of the profiles are shown in Fig. 5.7.

(b) Correlation of the north and south beach profiles

In order to correlate the amount of sand erosion and accretion with the relevant wind and wave data, an estimate of the amount of sand moved was attempted. Two successive profiles were superimposed on each other, and the area between them, representing the sand erosion or accretion along the line of survey, was measured by planimeter for both the north and south ends of the bay. These figures were plotted against the dates of consecutive surveys on two graphs (Fig. 5.2). The changes were normally in the opposite sense on the upper and lower beach. The points plotted on the graph show the change which occurred between two consecutive surveys. In general the curves, which show the correlation between the movement of sand at the north and south ends of the bay, agree reasonably well; although the amounts of sand moved differed, there were few occasions on which erosion took place in the north while accretion was dominant in the south and vice versa (Fig. 5.2). This is particularly well shown for the period 13 October–12 December 1950: the waves were then relatively large, and the changes recorded were on the whole greater than those of the summer period, 3 May–30 June 1950. The graphs in Fig. 5.2 show that, when erosion occurs on the upper beach, it is usual for deposition to take place on the lower beach. The reverse is also usually true. It is only on rare occasions that, over a fairly short period, erosion or deposition occurred throughout the whole beach profile, as it did at the north end between 2 and 12 December 1950 (Figs. 5.2 and 5.7).

It is noteworthy that the same type of beach change is nearly always recorded at both ends of the bay, suggesting that longshore movement in this small enclosed bay may be relatively restricted. This agrees with evidence derived from the drawing of wave refraction diagrams for the beach. These show that the longer waves approaching from any direction tend to turn parallel to the shore by the time they break, which would reduce the amount of sand moved alongshore by beach drifting.

(c) Correlation with tides

The weekly surveys were made at spring and neap tides alternately in order to ascertain whether there was a correlation between the state of the tide and the sand movement on the beach, such as that recorded from certain Californian beaches during some periods of the year. In California, work on the spawning on the Grunion (*Leuresthes tenuis*) (Clark, 1925) led to the observation of these changes, and the findings have been confirmed by the later work of Lafond (1939) and Shepard and Lafond (1940). It was found that, under normal calm conditions, the beach was eroded slightly above mean tide level at spring tide, while at neap tide deposition took place in the same zone. A similar close correspondence of tidal phase and sand movement was not recorded on the beach at Marsden Bay, although a slight swing of the beach with the tides was noticed on occasions, as, for example, at the north end of the bay for the profiles of 8 and 16 June (Fig. 5.7). At this time conditions were

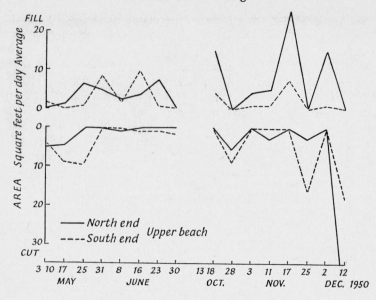

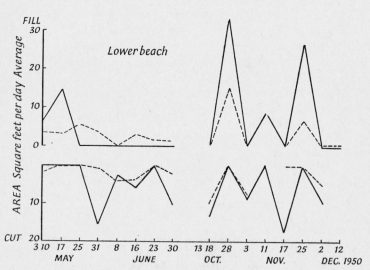

Fig. 5.2   *Graph to correlate the sand movement at the north and south ends of the bay. The top portion of the diagram is for the upper beach and the bottom portion for the lower beach. The continuous cut at both ends of the beach between 2 and 12 December is shown in the upper curve.*

relatively calm (Fig. 5.6): these are normally accompanied by constructive waves. But there is little evidence that this factor is of major importance at Marsden Bay, where long periods of calm weather seldom occur.

The cause of this sand movement is probably related mainly to the level at which constructive waves are most effective. Since this is at, or a little below, high-water level, the sand deposited and built up near the high-water level of neap tides is generally moved up the beach by spring tides to the new high-water level. Under these

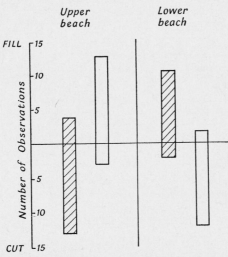

Fig. 5.3 Diagram to show the close correlation between wind direction and sand movement. Observations at both ends of the bay are combined.

conditions, therefore, erosion takes place near the neap tide high-water level at spring tides, and deposition at neap tides.

(d) Correlation with wind direction

A more important factor influencing the movement of sand is the wind direction. A graph (Fig. 5.3) was prepared by plotting the number of observations when erosion and accretion occurred on the upper and lower beaches respectively against the mean wind direction between surveys. This graph shows that there is a close correlation between wind direction and sand movement; accretion took place on the upper part of the beach with an offshore wind, and erosion on the lower part. The reverse applied with an onshore wind.

There are probably two reasons for this movement. First, the wind direction has an important effect on the waves reaching the coast. With an offshore wind the waves affecting the beach are normally long low swells which have travelled far from the generating area, while with onshore winds short, steep seas are generated. Wave steepnesses were therefore lower when an offshore wind had been blowing for a considerable period, as shown by the following figures, where the wave steepness is expressed as the wave height-to-length ratio:

Wind onshore		Wind offshore	
Maximum	0·0100	Maximum	0·0060 (very short waves)
Minimum	0·0020	Minimum	0·0002
Mean	0·0059	Mean	0·0014

The wave steepness has an important effect on the movement of beach material, since steep waves have been shown to be destructive on the upper beach, moving material to deeper water, while flat waves build up the top of the beach (Lafond, 1939; King and Williams, 1949). The maximum change normally takes place in the upper part of the beach, because here accumulates the coarser material, which leads to a steeper gradient so that the wave energy is dissipated over a smaller area than at low water, when the gradient is usually flatter.

Second, the wind has a more direct effect on the movement of sand by the operation of wind-generated currents, which are effective in moving sand thrown into suspension by the waves. An onshore wind blows the surface water shorewards, tending to create a seaward-acting bottom current; this results in a destructive effect near the shore in the shallow water. The reverse occurs with an offshore wind, but it is not so effective as an onshore wind, which is associated with steep seas and waves of greater energy.

Model experiments. The existence of such wind-generated currents has been demonstrated on a small scale by model wave-tank experiments. Measurements were made on the transport of sand under calm and windy conditions by measuring the amount of sand moved in a trap placed in different positions with respect to the break-point of the waves (King and Williams, 1949). In these experiments the waves were kept constant throughout; the results are shown in Fig. 5.4. The wave was constructive in type. Under calm conditions it

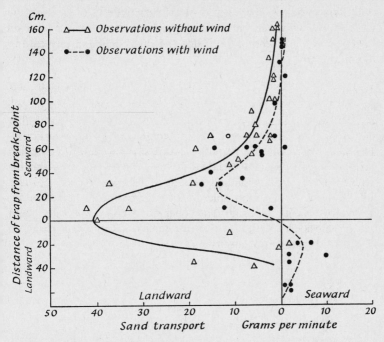

Fig. 5.4 Graph showing the effect of an onshore wind on the transport of sand both inside and outside the break-point. Wave height, 4 cm.; wave length, 350 cm. The original beach gradient was 1 in 15. Transport measured over the width of the tank (30 cm.).

moved sand landward in all depths, the amount increasing as the water became shallower towards the break-point, where a maximum was reached: this is shown by the full line in the graph. The effect of the onshore wind is shown by the broken curve. The landward movement of sand was reduced in all depths, but the significant fact is that this movement was reversed both in the shallow water landward of the break-point and in the deep water to seaward.

Fig. 5.5 shows two beach profiles built up by constructive waves of the same dimensions as those used for the experiment, the result of which is shown in Fig. 5.4. In the first profile a strong onshore wind was blowing and the resulting wind-generated undercurrent cancelled the constructive action of the waves so that the movement of sand was predominantly seaward. On the other profile, which shows the result of constructive waves under calm conditions, a marked accretion of sand on the upper part of the beach is shown.

These experiments demonstrate the destructive effect of an onshore wind in shallow water, particularly when it is accompanied by steep waves, which are destructive inside the break-point in any case in the model tank.

Marsden Bay. The surveyed profiles illustrate the same effect on a natural beach in many instances. To facilitate comparison of two consecutive profiles, these were plotted on the same graph so that the amount of erosion or accretion could be readily ascertained. The wind roses for the period between surveys were used to aid correlation of these two variables. Some of the wind roses and the profiles for the corresponding periods are shown in Figs. 5.6 and 5.7. In these surveys, examples of both erosion and accretion are shown. For clarity the conditions under which erosion was dominant will be considered first, and then the factors causing deposition on the upper beach will be discussed.

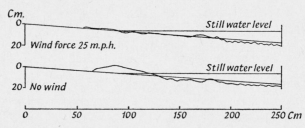

Fig. 5.5   Profiles of beaches formed in the model wave tank with constructive waves. Wave height, 4 cm.; wave length, 350 cm. The original gradient was 1 in 15. The profiles were drawn after 1½ hours. In the first experiment (top profile) an onshore wind of 25 m.p.h. was blowing. For the second experiment (bottom profile) there was no wind.

The period between 3 and 10 May was marked by north and north-east winds which resulted in erosion near the top of the beach, but not at the cliff-line itself in the south profile because the tide was neap during the period when the winds were blowing onshore. The most striking instance of erosion on the upper beach occurred between 18 and 28 October. The tide was between spring and neap, and strong north to north-east winds were blowing onshore. The waves during this period were steep and high (27 October, period 7 seconds, steepness 0·006; 28 October, period 6 seconds, steepness 0·010). They were locally generated short, high seas. The water-level at high tide was not sufficiently high to enable the waves to reach the cliff, so that the destructive action of the steep waves and onshore wind resulted in the cutting of a sand cliff 2 to 3 ft high throughout the greater length of the bay, while the material eroded from the top of the beach was deposited lower down, where accretion amounted to several feet of sand. Another example of marked erosion of the beach at the top occurred between 17 and 25 November, when the tide was rising to springs. Erosion during this period was particularly great at the south end of the bay, where 3½ ft of sand were moved

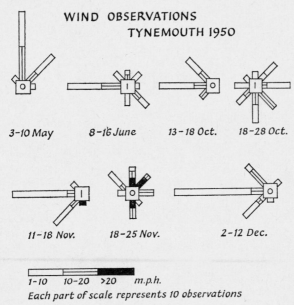

WIND OBSERVATIONS
TYNEMOUTH 1950

3-10 May   8-16 June   13-18 Oct.   18-28 Oct.

11-18 Nov.   18-25 Nov.   2-12 Dec.

1-10   10-20   >20   m.p.h.
Each part of scale represents 10 observations

Fig. 5.6   Wind roses for Tynemouth: six-hourly observations

from the upper part of the beach. Strong north to north-east winds blew during the week. Winds from this direction seem to be particularly effective at the south end of the bay because the north end is partially protected from this direction. Between 2 and 12 December erosion was again severe at both ends of the bay, and the high tide of the 12th enabled the waves to reach right to the cliffs. The waves were steep and high (period 9 seconds, steepness 0·0053) and the wind was mainly north-westerly. During the period 17 February–3 May 1950 the wind was blowing mainly from the north and north-west, which resulted in destructive wave action predominating; up to 4 ft of sand was removed from the north end of the bay over the greater part of the width of the beach.

The effects of offshore winds are shown in the profiles for 8 and 16 June, 13–18 October, 11–17 November, and 30 June–13 October 1950. The changes in the beach profile occurring during these periods

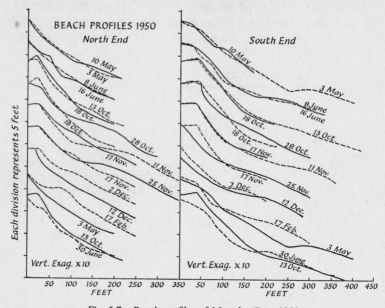

Fig. 5.7   *Beach profiles of Marsden Bay, 1950*

3–10 May: Onshore wind, neap tide on 10 May. 8–16 June: Offshore wind, low spring tide on 16 June. 13–18 October: Offshore wind, neap tide on 18 October. 18–28 October: Onshore wind, spring tide on 26 October. 11–17 November: Offshore wind, neap tide on 17 November. 17–25 November: Onshore wind, spring tide on 25 November. 2–12 December: North-west wind, high spring tide 12 December. 17 February–3 May: Mainly north and north-west wind. 30 June–12 October: Mainly south, south-west and west wind.

were similar; the upper beach was built up by the material moved shoreward from the lower beach, which was consequently eroded. The waves reaching the beach during such periods are long low swells; on 18 October, for instance, the wave period was 10 seconds and the deep-water wave height 0·4 ft. The waves were constructive in action. Rather similar waves, but of greater energy (period 11 seconds, deep-water height 0·8 ft), were acting on the beach between 11 and 17 November, causing accretion amounting to a maximum of $2\frac{1}{2}$ ft of sand at the north end of the bay. During the summer months from 30 June to 13 October the winds were on the whole light and offshore, with west, south-west and south winds predominating; considerable accretion took place throughout the bay, over the whole profile in the north and on the upper beach in the south.

(e) Beach gradient correlated with wave steepness and period

The beach gradient depends on the sand size, the wave steepness and the wave period or length. The sand of Marsden Bay is fairly coarse, and this accounts for the relatively steep beach profile. The rate of percolation through coarse material is more rapid than through fine material. This means that the backwash is rendered less effective; the force of the swash and backwash must be equalised by an increase of gradient, which increases the efficiency of the backwash but decreases that of the swash. The steepness of the beach causes a concentration of wave energy within a fairly narrow horizontal zone; this increases the mobility of the beach and explains the rapid changes in beach profile recorded.

The sand size is fairly constant on any one beach and thus the variation of gradient recorded must be accounted for by changes in the wave steepness and period. For waves of a constant period it was found that the gradient increased with the decreasing steepness of the waves. The gradient measured is that of the high-tide swash slope, which adapts itself most rapidly to the prevailing wave conditions, so that it may be assumed that the gradient is adjusted to the dimensions of the waves acting on it at high tide. The following hypothetical case may help to explain the process. It is assumed that the rate of percolation through the swash slope remains constant. As the wave steepness increases, the volume of swash moving up the beach will also increase. But since the amount of the swash percolating through the beach remains constant, the backwash will increase in proportion to the swash. If, for example, in a flat wave at any point half the swash is lost by percolation, then the backwash will be half the swash in volume. If subsequently the volume of swash is doubled, then only one-quarter of it will be lost by percolation and the backwash will be three-quarters of the swash. This relatively powerful backwash needs a less steep gradient to reach equilibrium.

Wave period, which is related to wave length, also affects the gradient of the swash slope, as illustrated in Fig. 5.8. The scatter of points is partly explained by the differing wave steepness involved.

ICD E

It was found possible to analyse the data statistically as the relationship between the beach gradient and the wave period was approximately linear. A significant correlation was found between the two variables, *r*, the coefficient of correlation, being 0·560; the probability of this result occurring by chance is less than 0·001. The regression lines, entered on the graph, show that as the wave period increases, so the beach gradient decreases. This result may be explained by similar reasoning to that already suggested. The backwash will again be increased in proportion to the swash.

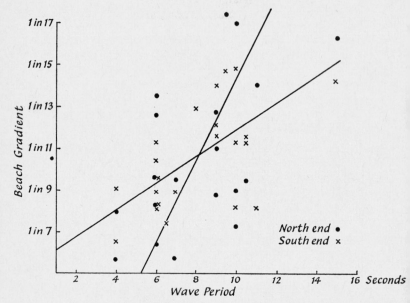

Fig. 5.8   Graph to show the correlation between beach gradient and wave period. The data have been analysed statistically; the coefficient of correlation is 0·560. The regression lines are drawn on the graph Regression equations:

$$X = 0·50Y + 16·64$$
$$Y = 0·63X + 9·13$$

Both these correlations have been demonstrated in a model wave tank, and it is interesting to establish that they can be applied to beaches in nature and that the results can be checked by statistical methods to confirm their significance.

CONCLUSIONS

The observations and surveys made at Marsden Bay have shown that the sand moves predominantly normal to the coast and that longshore drift is limited. This feature can be explained by the nature of the coastline and the refraction suffered by the long waves which usually reach the beach.

The correlation between the direction of movement of beach material, wind direction and wave type agreed well with results obtained in a small wave tank. This correlation is therefore probably of wider application. The gradient of the beach was found to vary with the wave steepness and period in a way similar to the swash gradient under controlled conditions of wave-tank experiments. These results also seem to have a wider application and show that results of model experiments can have a useful application to natural phenomena. The number of variables involved in the movement of sand on a natural beach complicate the issue, but it has been shown that both wind direction and wave type are of fundamental importance and that these two factors are closely linked together.

The findings of the analysis of the observations, with regard to changes in beach profile to be expected under certain well-defined conditions, are summarised below:

(i) *Swell* – long waves, *onshore* wind – high waves.
Waves are fairly steep and their energy is very great. These waves are markedly destructive, moving sand seaward over a considerable width of beach. The resultant gradient should be flat owing to the long period of the waves; this, combined with the steepness, should give a flat gradient over a fairly wide zone.

(ii) *Swell* – long waves, *offshore* wind – low waves.
Waves are very flat and their energy is moderate. These waves are constructive in type and move sand up from the lower to the upper beach in considerable quantity as their energy is great. The gradient should be fairly flat under the influence of the long waves, which are the controlling factor in determining gradient.

(iii) *Local sea* – short waves, *onshore* wind – high waves.
Waves are very steep and their energy is moderate. These waves are very destructive at the top of the beach and move sand to the lower beach. Gradients tend to be fairly flat as

in this case the steepness of the waves should be the dominant factor.

(iv) *Local sea* – short waves, *offshore* wind – long waves.

Waves are very flat and their energy is very small. These waves are constructive but their action is limited to a narrow zone. In this zone the gradient is steep, as both the short wave period and flat waves tend to produce steep gradients.

There are many more combinations of the factors affecting sand movement, but the ones listed indicate some of the possibilities.

The conclusion may be reached that on Marsden beach the prevailing offshore wind is associated with constructive wave action while the dominant onshore wind has a destructive effect.

REFERENCES

BEACH EROSION BOARD (1942) 'A summary of the theory of oscillatory waves', Technical Report No. 1 (Washington).

CLARK, F. N. (1925) 'Life history of *Leuresthes tenuis*', State of California Fish and Game Commission, Fish Bulletin No. 10.

KING, C. A. M., and WILLIAMS, W. W. (1949) 'The formation and movement of sand bars by wave action', *Geogr. J.*, CIII 70–85.

LAFOND, E. C. (1939) 'Sand movements near the beach in relation to tides and waves', *Proc. VIth Pacific Science Congress*, II 795–79.

SHEPARD, F. P. (1950) 'Beach cycles in Southern California', Technical Memorandum No. 20 (Beach Erosion Board, Washington) 1–26.

—— and LAFOND, E. C. (1940) 'Sand movements along Scripps Institute Pier', *Amer. J. Sci.*, CCXXXVIII 273–88.

6 The Influence of Rock Structures on Coastline and Cliff Development around Tintagel, North Cornwall

GILBERT WILSON

ABSTRACT

The paper deals with the morphology of the stretch of coast between Boscastle and the Treligga Cliffs in north Cornwall. The prevalent cliff type is slope-over-wall, either hog's-back or bevelled as defined by Miss Arber (1949). Flat-topped cliffs are less common. It is considered that geological structures – bedding, jointing, faulting and, to a lesser extent, rock types – have not only had considerable influence on the shape of the cliffs, but have also guided the marine attack on the coast. The evolution of most of the coast features, when seen in plan or profile, has been controlled by the local geological structure. The southern stretch of the coast is sub-parallel to seaward-dipping normal faults which have been stripped of their hanging walls. Such a cliff-line is dominantly a fault-line scarp which has suffered slight retrogression.

1. INTRODUCTION

THE control of coastal morphology by the lithology of the rocks exposed to wave action has long been recognised by workers in north Cornwall. Thus Dewey (1914), in his description of this region, written preparatory to the Association's excursion to the district, stated (p. 155): 'The varying degrees of hardness of the rocks, which play only minor parts in producing the topography of the inland localities, are the main cause of the coast scenery. When the rocks are exposed to the violence of the sea, the soft killas is rapidly destroyed and washed away and wide bays result; but the sills and lava offer resistant masses which protrude into the sea as great headlands. . . .' Dewey lists among these promontories those of Trevose, Stepper, Com, Varley Head and Pentire, 'a towering mass of pillow lava with vertical cliffs rising majestically from the sea to a height of nearly 300 ft'.

The active erosion that this passage suggests is now disputed by more recent workers. Thus Balchin (1946, p. 341) considered that there had been 'small change in the morphology of the area throughout later Pleistocene and Recent time: both sub-aerial and submarine denudation are alike in this respect . . . the location of the coastline has probably altered but little even if the detailed morphology has been subject to modification'. This statement is contained in a paper in which the morphology of north Cornwall was discussed from a regional standpoint. In the present article, however, it is the local modifications of the coast that are considered.

In many localities in the area it has been observed that active erosion is proceeding along certain privileged paths which are structurally controlled. Such paths, in the case of flat or gently dipping structural planes of weakness in the rocks, are important only when the level of the sea happens to agree with that of the structural feature – literally a case of coincidence – and it is only rarely that such features are of much significance. Steeply inclined planes of weakness are, however, important over a big range of sea-level. They are paths of easy attack on the coast whether the waves break at 50 or 100 ft above, or 50 or 100 ft below, the present tidal range. It is therefore reasonable to assume that the same relationship, which is seen today between the planes of weakness and marine erosional activity, also held in the past when the sea-level was different. It is on the basis of this assumption that the accompanying thesis has been developed.

The section of coastline discussed in this paper runs from Pentargon and Boscastle in the north, to Backways Cove and the Treligga Cliffs in the south (Fig. 6.1), a stretch of about six miles. From Boscastle to Tintagel the coast trends approximately S.60°W., from Tintagel to Backways Cove the trend is a few degrees west of south. Except for the embayment of the Trebarwith Strand beach, the latter section presents a much smoother plan than does the more crenulated strip between Tintagel and Boscastle. The cliffs are cut into the 300–430-ft Trevena Platform of Pliocene age described by Balchin (1937, 1946), and in many places drop sheer from the platform to below sea-level. The number of routes by which one can reach the sea is distinctly limited. Over much of the coast the cliffs show a *slope-over-wall* form (Whitaker, 1909), in which a vertical drop lies below a grassy slope which may be curved (convex upwards)

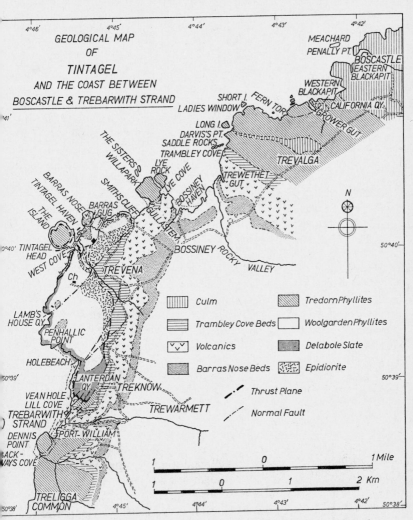

Fig. 6.1 *Geological map of the coastal area around Tintagel from Boscastle to Treligga Common. [Reproduced from Wilson (1951) by permission of the Council of the Geological Society.]*

thus forming a *bevelled cliff*, or may be a uniformly inclined surface yielding a *hog's-back*. The origin of these forms, which will be considered later, has been discussed by Balchin (1946) and Miss M. A. Arber (1949). Similar forms have been described from the Devonshire coast by E. A. N. Arber in 1911, and from Wales by Challinor (1931). Recently Cotton (1951) has proposed a two-cycle process, in which glacial control of sea-level plays a prominent part, for the evolution of the *slope-over-wall* cliffs of south-west England. The Cornish coast and other aspects of the geomorphology of south-west England have also been studied by Professor André Guilcher of Paris (1949, 1950), who compares the various features observed with those seen in Armorica.

The present writer, during periods of investigation of the tectonics of the Tintagel area (Wilson, 1951), became more and more impressed by the importance of structural features in influencing the local evolution of the coast and by the comparatively minor role of lithology. In the pages which follow, the part played by structures in the rocks and the manner in which they control the cliff morphology is discussed, and an endeavour is made to picture the principles of wave attack on this picturesque piece of coastline.

The writer wishes to thank Professor H. H. Read for suggestions made during the writing of this paper, and also to express his gratitude to Mr G. S. Sweeting, who gave much helpful advice and constructive criticism. The map (Fig. 6.1) has been reproduced from Wilson (1951) by kind permission of the Council of the Geological Society of London, to whom thanks are also tendered for the use of the original block.

2. GEOLOGICAL STRUCTURE

The current interpretation of the rock structure in the Tintagel area was formulated by Dewey (1909) and was later summarised by him in 1914 and 1935. Recent investigations have shown, however, that the tectonic history of the area was more complicated than originally believed (Wilson, 1951). As Dewey so convincingly demonstrated, the coastal strip from Trebarwith Strand to Smith's Cliff is built up of overthrust slices of Upper Devonian strata. Recent mapping has shown that the movements which drove these slices into their present positions did *not* come from the west-north-west as believed, but came from the south-south-east towards the north-north-west. The piled-up strata were then broken by roughly parallel normal faults, with downthrows towards the west and north-west, and were also warped by the gentle fold of the 'Davidstow Anticline'. The normal faulting is a factor which has had considerable influence on the forms of the cliffs in the Tintagel–Trebarwith Strand area.

The strata along the coast from Boscastle in the north, at Tintagel, and thence southwards to Treligga Common, south of Trebarwith Strand, are of Lower Carboniferous and Upper Devonian age. The latter beds are altered by regional metamorphism to low-grade phyllites in which chloritoid and ottrelite locally occur. The Carboniferous rocks are Lower Culm and show little sign of metamorphism, although they are much contorted. They have been described by Owen (1934, 1950). The general succession, youngest beds at the top, is as follows (Dewey, 1909):

Carboniferous

Lower Culm	Black and grey carbonaceous shales with grit bands. A greenish volcanic band locally occurs at the base.

〰〰〰〰〰〰〰 Unconformity.

Upper Devonian (with *Spirifer verneuli*)

Tredorn Phyllites	Pale-coloured, felspathic phyllites, locally slate.
Trambley Cove Beds	Dark blue-black soft slates and siliceous grits, often much broken.
Volcanic Series	Pillow lavas and tuffs, now altered to chlorite-schists, with epidote, calcite and biotite. A thin limestone band is locally found at the base.
Barras Nose Beds	Dark blue-black slates and siliceous grits, often much broken.
Woolgarden Phyllites	Thin banded greenish-grey micaceous phyllites, with chloritoid and ottrelite.
Delabole Slates	Grey-blue or grey-green slate, with rare grit bands.
Epidiorite	Intrusive. Locally massive, but commonly sheared to green chlorite-schist. Occurs in the Woolgarden Phyllites at Tintagel.

The succession from the Culm to the Woolgarden Phyllites is exposed along the cliffs between Boscastle and Bossiney Haven. The

ICD E 2

general dip is northerly, as one is here on the south flank of the great Culm Syncline of Devonshire. The strike of the beds changes at Bossiney Haven, where it swings from roughly east–west to north–east and thence to north–south. The dip of the beds changes correspondingly from northerly to westerly. The fold responsible for this change in orientation of the beds is the Davidstow Anticline (or hemi-dome), the axis of which passes through Bossiney Haven and plunges gently to the north-west (De la Beche, 1839, p. 56; Pattinson, 1847, p. 8; Parkinson, 1903). These beds form the unmoved foundation rocks of the area, and from Smith's Cliff to Trebarwith Strand they are overlain or faulted against the thrust-slices which lie along this coastal stretch.

The rocks of the overthrust region have been grouped into the following tectonic units (or thrust-slices) separated by three thrust-planes: T_1, or the 'Flat Thrust', T_2 and T_3. From west to east these comprise:

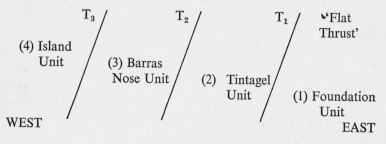

The more westerly or higher tectonic units have been preserved from erosion by the general dip to the west and by normal faulting which throws down the thrust-slices to the west or north-west (Fig. 6.2). The thrusting movements took place along planes which were originally more or less flat, but owe their present inclination, which carries them below the sea, to later movements. The direction of translation was towards the north-north-west, and is shown by the orientation of various minor structures. These include:

1. *Elongation.* Slickenside striations, stretching of lava pillows, ejecta, amygdules, etc.
2. *Tension structures.* Tension fractures and boudinage oriented normal to the direction of elongation.
3. *Drag-folding and Cleavage.* Fold axes are normal to the direction of movement, axial planes and cleavage dip to the south-south-east.

The main movements were of course along the thrust-planes, but there was also slip throughout the pile of phyllites analogous to irregular sliding of a pack of cards, and 'the motion took place most readily along the direction of bedding' (Parkinson, 1903, p. 410). Consequently the minor structures or tectonic weathercocks of the types mentioned are not confined to the shear zones, but may occur in any exposure over the whole area.

Around Boscastle, in the Culm beds, zigzag folding is predominant. The general dip is to the north, but the folds indicate a

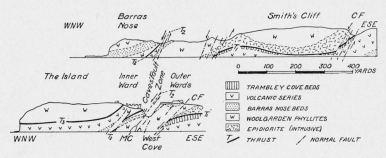

Fig. 6.2 *Generalised geological cross-sections at Tintagel*
CF= Castle Fault.   MC= Merlin's Cave.

down-dip movement, that is, the upper beds moved north-north-westerly relative to those below (Owen, 1934, pl. 41b; Fourmarier, 1936, p. 1032).

In the Devonian beds drag-folds are less common than were the zigzags in the Culm, but their axial trends and the direction of overturning are similar. Stretching structures, such as boudinage, are also parallel to these axes of folding. Elongation structures are best seen in the volcanic rocks on Trebarwith Strand, where lava pillows and ejecta are drawn out in the direction of movement and trend north-north-west at right-angles to the drag-fold axes (Wilson, 1951, Fig. 10).

The movements on the normal faults, which throw down to the west and north-west, were deduced by similar structures. These include drag of the beds on either side of the fault-planes, the formation and orientations of tension fractures, and in some cases drag-folding, all of which indicated a down-dip movement (Wilson, 1951, pl. 30, Fig. 13). The important fault zones have been given local names: the *Castle Fault* runs from West Cove, up past the Castle

Gateway, across the valley of the stream from Trevena, and thence
to the east end of Smith's Cliff. The *Caves Fault Zone* forms the
spectacular zone of faulting which runs up the cliffs of The Island,
thence it crosses Tintagel Haven and passes northwards out to sea
at Barras Gug (West). Other normal faults lie between these two
main zones. The sea cliffs both to the north and south of Tintagel

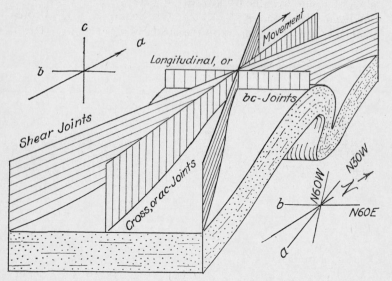

*Fig. 6.3 The main directions of jointing in relation to the folding. The three
structural axes, a, b and c, are indicated, and the orientation of the diagram, in
respect to the Tintagel area, is shown in the lower right-hand corner.*

are in places cut by similar faults which are all members of the same
group.

The importance of these two movements – the thrusting and the
normal faulting – from a morphological point of view, is their production of planes of weakness in the rocks. The thrusts are more or
less flat and are not always well marked, but the normal faults have
formed sloping shear zones along which wave action at any level can
do much damage. Additional planes of weakness in the form of joint
systems have resulted from both tectonic movements, and these too
have been utilised by the wave attack on the cliffs.

Although joints may form in rocks on almost any bearing, there
are certain preferred directions, relative to the direction of the movement which has taken place, that can be recognised as being optimum

joint orientations (Cloos, 1937; Willis and Willis, 1934). These are shown in Fig. 6.3, in which the three structural axes, *a*, *b* and *c*, are also indicated (Wilson, 1946, p. 265 n.; Cloos, 1937): *a* is the direction in which the movement took place; it is that of translation; *b* is normal to *a*, and is the direction about which rotation structures due to the movement would develop. It is recognised in the field by the trends of fold axes and commonly by minor corrugations. Here it is also shown in addition by the strikes of tension fractures and the elongation directions of boudins; *c* is normal to *a* and *b*, and in this area it is approximately vertical.

The relationship of the important joint directions to the movement is illustrated in Fig. 6.3, and it will be seen that the joints occur in two main systems. The first, comprising dip- and strike-joints, are those in the *ac*-plane (*cross-joints*) and those which are parallel to *b* (*longitudinal joints*) respectively. The former are common in most folded regions. Strike or longitudinal joints are often inclined, and their orientation may be influenced by cleavage directions, etc. Some are filled with quartz or calcite, and others gape open as a result of tensional drag due to continued local movement of the rocks. Quartz-filled fractures are well seen in the cliffs between Merlin's Cave and Tintagel Head and in the epidiorite at the foot of the waterfall in Tintagel Haven. The two other joint sets form *oblique joints*, complementary to each other, and are more or less symmetrically arranged on either side of the direction of movement *a*. They commonly make an angle of less than 45° with the *a*-axis. These are shear-joints, and their relationship to the rock deformation and the other structures is seen in small scale among the elongated pillows of Trebarwith Strand (Fig. 6.4).

It is unusual for all these joint directions to be equally well developed over the whole area; the formation of one or more sets of joints may prevent the development of others by relieving the stresses responsible.

The joints of the Tintagel area, which are related to the overthrust movements, result from a drive towards N.30°–35°W. Vertical cross-joints striking in that general direction are well developed; *bc*-joints at right-angles to this direction are locally important. Oblique joints are nearly as common as cross-joints; one set only, however, has developed almost to the exclusion of the other. North–south joints, which strike 30° to 35° east of the direction of movement, are met with throughout the area. The joints, which represent the complementary shear direction, about N.60°–70°W., are relatively rare.

Joints which are connected with the normal faulting movements and show no relation to the earlier thrust structures are locally of importance. They lie roughly parallel to the dislocations, strike north-east to south-west and dip about 45° to the north-west.

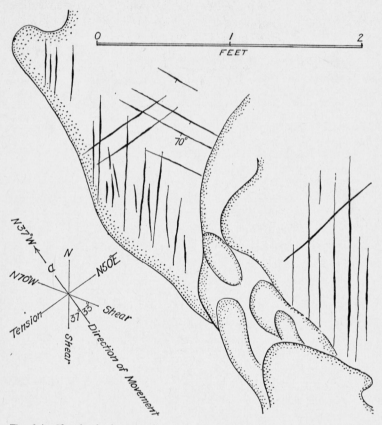

Fig. 6.4 Sketch of a horizontal rock-surface, showing joint-patterns in deformed pillow-lava and their relation to the movement directions. Trebarwith Strand.

The influence of these many structural planes of weakness cannot be neglected when the local evolution of the coastline is considered. All of them, dip and strike of bedding, thrusting, faulting and jointing, as well as rock type, have played their part to a greater or lesser degree in guiding the agents of erosion, and share the responsibility for the development of the spectacular cliffs of this coast.

3. STRUCTURE AND CLIFF-FORM

It is generally agreed that erosion by unarmed waves on a stretch of cliff coast in tough homogeneous rock is slow, and such cliffs may be fairly old. Where the rock is not homogeneous, however, wave action will, as soon as planes or zones of weakness are exposed, break up a smooth and orderly rocky coast and will develop local crenulations. Thus resistant ramparts become indented by narrow joint-controlled inlets, locally termed 'guts' or 'gugs'; fault zones are eroded into chasms or caves; and landslips develop on undercut seaward-dipping bedding planes and inclined fractures. By means of narrow channels, wave action can penetrate into more easily eroded zones lying behind the resistant rocks which formed the original cliff-line. Once soft beds or a fault zone of crushed rock are reached in this way, the attack proceeds laterally along them; if these are sloping, the rocks are undercut and falls or slides are promoted.[1] Narrow inlets more or less parallel to the coast are thus formed.[1] The joining together of such inlets leads to the development of stacks, which, once isolated, are 'mopped up' in due course.

Downward corrosion by streams also tends to breach the regular cliffline. The sides of estuaries form funnel-shaped openings into which pile seas that arm themselves with alluvial detritus. Scour and erosion of that side of the estuary which faces towards the more powerful storm winds are thus promoted, and again zones of weakness behind the main cliffs may come under the influence of marine attack.

All these factors have played parts of varying importance in guiding the wave attack on this coast. In some cases two or three factors have combined, and the picturesque scenery of the cliffs is largely dependent on the variety of structural guides available. This can best be illustrated by considering various stretches of the coast and their relationships to the local geological structures.

The coast immediately north of Boscastle, between Penally Point and the inlet of Pentargon, is formed of beds which dip northerly (Fig. 6.5). The cliffs here are recognised as having typical hog's-back form, and their upper, even, slope terminates in the flat cultivated surface of the Trevena Platform. Downwards, the grassy incline ends abruptly in vertical and locally undercut sea cliffs. The slope closely

---

[1] The writer has observed inlets formed in this way among the 'geos' of Farr Point, Bettyhill, Sutherland. Other better-known examples occur at Lulworth Cove, Dorset.

corresponds to the general angle of dip of the beds on the south side of the inlet, but the north shore is a steep scarp type of cliff. Within Pentargon itself is one of the few local wave-cut platforms which correspond with the present sea-level, and above this area, where active erosion is taking place, the ground to the south is slipping seawards. The more central part of this northward-facing coast has an appearance of stability. The hog's-back slope seems to be a continuation of the valley of the stream which flows into the sea at the head of Pentargon. This part has not been cut back to the same

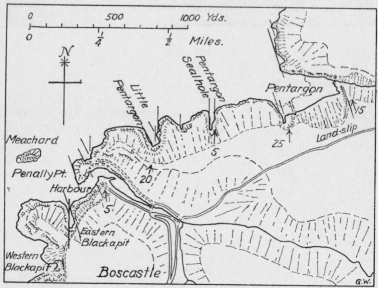

Fig. 6.5 Map of the coast around Boscastle and Pentargon, showing the Pentargon hog's-back cliff and the strike-directions of local steeply dipping joints

extent as the present landslipping portion, because the relatively deep water below the cliffs did not permit the waves to arm themselves with rock debris, thus slowing down erosion.

The importance of jointing in the development of steep-sided narrow guts is well exemplified at Pentargon (Balchin, 1946). In addition, the cliff at the head of the inlet itself is largely a joint-plane striking N.25°W. – a cross- or *ac*-joint – and the points of the headlands north of the bay are truncated by similar planes. Pentargon Sealhole, the cliffs of which overhang a wave-cut nip, is eroded along north-south joints. The shape of Little Pentargon is controlled by intersecting north-south and N.20°W. joints.

The estuary of the Valency at Boscastle is a drowned valley, but evidence that the sea-level may have been higher than at present is suggested by a prominent little wave-cut nip in the cliffs above the present high-water mark. The rocks one either side of the river mouth are strongly jointed, and the end of Penally Point is cut off by fractures striking N.20°W. Similar joints guided the erosion of Eastern Blackapit.

The Boscastle Blow-Hole which passes through the neck of land joining Penally Point to the mainland is also eroded along such joint-planes. At low and mid-tides the south exit of the tunnel can be seen, and when a big sea is running ouside great jets of water

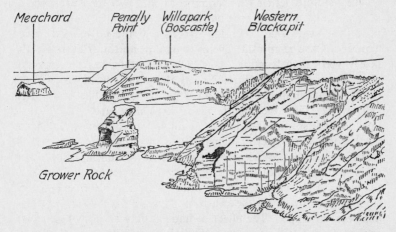

Fig. 6.6   Hog's-back cliffs between Boscastle Old Quarry and Boscastle, looking north-easterly. (From a photograph.)

are thrown into Boscastle Harbour. The mouth of the hole is covered at high tide. Anyone who has seen this blow-hole belching forth its great gushes of water and spray cannot but appreciate the power of waves to erode by compression of air or water in joints and rock crevices.

South of Boscastle, erosion along joint-planes leads to many irregularities in the coast. Grower Gut is an example of a rectangular inlet which has been cut back along west-north-westerly or oblique joints. The fault, which can be seen at its head, continues northwards and a narrow cave has been eroded along it near California Quarry.

The sloping hog's-back cliff faces between Grower Gut and Boscastle Old Quarry (Fig. 6.6) are controlled by joints, and in places by minor normal faults, which dip about 45° to the north-west. A second set of joints striking N.30° to 40°W. intersects the former, and erosion along the two sets has led to the local development of a rectangular shore pattern. The stretch of cliff, which runs north-west from Boscastle Old Quarry to Short Island, is another hog's-back, here controlled by the dip of the bedding. Wave attack along steeply dipping joints has resulted in the formation of small stacks and tidal reefs, as well as vertical gullies which cut the cliffs.

Short and Long Islands are great stacks rising to the level of the Trevena Platform (Dewey, 1914, Pl. 2A), and are separated from the mainland by erosion along joints. In this they differ from Saddle Rocks which, together with Trambley Cove, owe their existence to normal faults. Erosion here has attacked along a fault zone which forms the south side of Saddle Rocks and the north side of the Trambley Cove re-entrant. The bedding dips northward towards the nearly vertical dislocation. Trambley Cove thus has a wedge-like form: the north side has broken away along the fault-planes, and the south side has slipped down the dip slope of the Trambley Cove Beds.

Trewethet Gut, the mouth of Rocky Valley, and a precipitous inlet, 200 yds south of the latter, are all largely controlled by vertical north-north-westerly cross-joints. North and south of Rocky Valley, however, inclined joints associated with minor normal faults and dipping about 45° north-westerly exercise considerable influence on the forms of the cliffs. Where the attack has laid bare these planes, a sloping glacis up which the swash of the waves largely expends itself has been developed. The effect of undercutting along such inclined surfaces leads to the occurrence of overhanging cliffs and to rockfalls, with the eventual production of an evenly sloping cliff.

The walls of Bossiney Haven are largely vertical joint-planes, and enlargement of such fractures has produced the rock pillars which form the well-known Elephant Rock. There is a marked swing in the strike of the beds about here, and the bay of Bossiney Haven is eroded in the nose of the gently plunging Davidstow Anticline.

Between Bossiney Haven and Lye Cove is a rounded headland composed of altered volcanic rocks, now chloritic schists. This is one of the few local examples of such rocks forming a promontory. It is, however, noteworthy that on the opposite side of the bay the volcanic rocks do not produce any corresponding feature.

The promontory of Willapark, Lye Rock and The Sisters is composed of Tredorn Phyllites which are much jointed. The great chasm between Willapark and Lye Rock is eroded along N.30°W.

or *ac*-joint-planes; and the vertical flat-topped cliffs on the south-west side of the big promontory have developed on shear joint-planes striking N.60° to–65°W., and *ac*-joints at N.30°W. Inclined jointing, which dips north-westerly, also plays a part in the form of Willapark. Such joint-planes can be seen cutting across the face of Lye Rock, and the sloping north-west faces of both Lye Rock and of Willapark are largely controlled by these fractures. This seaward face of the promontory presents an excellent example of a hog's-back. The vertical cliff at the foot of the slope is about 50 ft high, and the remaining 250 ft to the top is taken up by one even, structurally controlled, sloping surface. The gap between Willapark and The Sisters, as well as that between the two stacks, is eroded along joints. Balchin (1946) points out how commonly erosion along joints has led to the formation of stacks in this partciular type of rock – the Tredorn Phyllite – and certainly one sees from the map the reason for this statement.

The hog's-back cliff-line which can be followed from Lye Cove, across the isthmus of Willapark to Gullastem, and thence to the re-entrant at the east end of Smith's Cliff, is formed by a combination of dip-slope and normal faulting. The soft Trambley Cove Beds, which are lying on the top of the Volcanic Series, are broken by normal step-faults. The bedding and the faults both dip seawards, i.e. towards the north-west (Fig. 6.7). The dual effects of the dip of the bedding and of the fault-planes have been to provide sloping slip-surfaces from which the overlying and softer rocks were peeled off. The top of the Volcanic Series forms the present slope, and, combined with the marine attack at the foot, it has produced the present slope-over-wall cliff. Wave erosion along steeply dipping joints is locally breaching the face of volcanic rocks and forms rectangular re-entrants in the cliff. Continuation of attack along these planes will eventually penetrate into the less resistant Barras Nose Beds below the Volcanic Series; the direction of excavation should then turn along the strike of the strata.

Between Smith's Cliff and Barras Nose the geological structure becomes more complex than hitherto. The Castle Fault is responsible for the appearance of the Flat Thrust (T_1) near the foot of the cliff. The thrust zone is locally eroded into a horizontal saw-cut above which the cliffs overhang. The sole of the thrust is a flat ledge of altered volcanic rock partly covered by rock-fall and slide debris. West of Smith's Cliff it is found that cross-joints striking N.35°W. become common, and inlets with vertical or undercut walls having this orientation have been eroded. The cliff between East and West Barras Gugs is seamed with enlarged vertical *ac*-joints, but the two 'gugs' themselves are the result of erosion on normal faults (Fig. 6.8).

The faults of East and West Barras Gugs strike roughly at right-angles to the coast, and the two inlets, though steep-sided, are asymmetrical. They can both be descended (with difficulty) on their

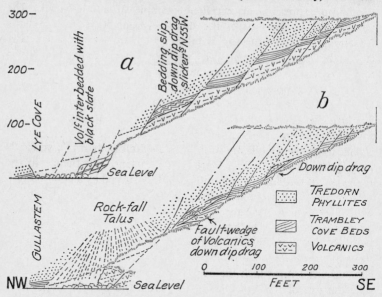

Fig. 6.7 *Sections showing the relationship between the hog's-back cliff on either side of the Willapark isthmus and the geological structure. (a) Section to the north-east; (b) Section to the south-west of the isthmus.*

sloping eastern surfaces formed by the stripped fault-planes and down-dragged beds. Their western walls are vertical or overhanging.

The fault in West Barras Gug continues through the isthmus and is eroded into a precipitous chasm ending in a cave. From here this line of dislocation – the Caves Fault Zone – splits, and is responsible for the group of faults along which caves have been eroded on both sides of Tintagel Haven. Two of these caves are 'through caves'. One is a low tunnel passing beneath the rectangular promontory of Blackarock; the other is Merlin's Cave which passes below the ruins of the Castle on The Island from the Haven to the corner of West Cove. In rough weather with a south-westerly gale the seas, piling through Merlin's Cave into the Haven, form a very impressive spectacle.

The Caves Fault Zone played an important role in the later development of Tintagel Haven. This inlet, separating The Island from Barras Nose, is the estuary of the Trevena stream which now cataracts down some 40 to 50 ft on to the bay-head beach. The former continuation of this stream and its valley can still to a certain extent be extrapolated (Fig. 6.8). The levels of the top of the Blackarock

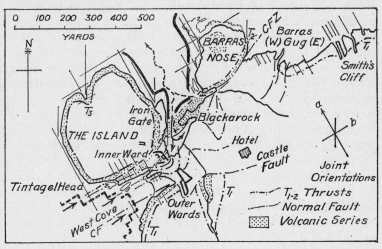

Fig. 6.8   *The evolution of Tintagel Haven. The first stage is shown by a heavy line; the directions of the second stage of the wave-attack are indicated by a branching arrow. Merlin's Cave is shown passing below the Inner Ward. Thin lines represent prominent joint-directions.*

CFZ = Caves Fault Zone.   CF = Castle Fault.

promontory and of the projecting step upon which stands the Iron Gate would roughly coincide with the existing valley floor if it were prolonged seawards at its present gradient, and the grass and rock slope, rising above the Iron Gate on The Island, is accordant with the south-west side of the present Trevena valley. A one-time tributary of this prolonged river is seen in the smaller brook which now cascades into the sea between Barras Nose and Blackarock. It is evident that this line of drainage has had considerable influence on guiding the marine attack inland, and was responsible for the initiation of the excavation of the Haven.

The base of the seaward cliffs of The Island is composed of sheared volcanic rocks. Similar rocks occur as an outlier on Barras Nose and dip gently to the north-west. It seems probable, therefore, that the original coastline, from which this portion of the present shore has been carved, lay west and north-west of these two promontories, and consisted largely of the relatively tough Volcanic Series.

The mouth of the prolonged Trevena stream would make a re-entrant in this shoreline, and would also be responsible for the thinning or possibly the cutting through to sea-level of the volcanic facing. The stream mouth would thus assist in exposing to wave attack that portion of the coast where the hard rock-band was thinnest or even broken. North-north-westerly (cross) and north–south (oblique) jointing would guide the first stages of the attack inland, and blocks would tend to break away along these and east-north-east (or *bc*) joints. Once the outer north-westerly-dipping volcanic band had been penetrated, it must have been undermined by the erosion of the softer Barras Nose Beds below it. This would be particularly so on the Barras Nose side of the Haven where the dip is steeper and the rocks are more exposed to the full force of westerly gales. Here is another place where a wave-cut platform corresponding to the present sea-level occurs.

The shape of the outer part of the Haven agrees with the general joint-pattern, by which its excavation was undoubtedly controlled (Fig. 6.8). The pattern is simple on the south-west side of the Haven, but is more elaborate on the east or Barras Nose side. This joint-control seems to have continued as far as the line between the Iron Gate and the south-west corner of Barras Nose. Here the presence of the tougher Woolgarden Phyllites made itself felt, and the Haven narrows markedly; but the trend of the Barras Nose cliff still reflects the importance of north-north-west jointing. South-east of and below the Woolgarden Phyllites is the thrust (T_2), a tight dislocation which does not act as a significant zone of weakness. It separates the Phyllites from an underlying band of volcanic rocks. This band of tough, but well-jointed, chlorite-schist formed a natural barrier protecting the head of the bay, even as similar rocks now do along the coast between Willapark and Smith's Cliff.

Eventually this protective wall was broken through, but its abutments can be recognised in Barras Nose, Blackarock and The Island. One can picture the manner in which it was pierced. The waves would be attacking a sloping rock face traversed by joints, from which blocks would periodically be displaced. During a heavy sea the water would be spouting from these fractures, even as it does on Blackarock today. Massive joint-bounded blocks would thus be wrenched away from their setting and would tend to slip seawards down the dip-slope, in a manner similar to that seen at present at Trebarwith Strand. Once a breach had been opened through the resistant rock barrier the waves would encounter the underlying soft and shattered rocks of the Barras Nose Beds and the Caves Fault Zone. The attack would then fan outwards on one or both sides of the breach, as shown by branching arrows in Fig. 6.8. The immediate result of this breakthrough was the removal of support from behind the seaward-dipping volcanic rocks which would thereupon collapse and be broken up piece by piece. At the same time the waves, armed with rock fragments, would undermine the lower part of the foot-wall slope of the Caves Fault Zone, thus causing further slip of the hanging-wall rocks down the fault-planes which here dip at about 35° to 45°. Such slip has been responsible for the formation of the long-sloping fault-controlled hog's-back which rises from the present cliffs up to the 300 ft platform near the Hotel. It seems probable that the caves were driven into the fault zone within a short time of the gap being formed. The waves surging through the breach into a restricted space would, in the early stages, have greater tunnelling power than later when the bay was more fully developed.

Apart from the main break leading into the head of Tintagel Haven, a second gap was also formed between Blackarock and Barras Nose, on the line of the little tributary stream. The isthmus connecting Barras Nose with the mainland is now being tunnelled along the fault zone in a manner similar to that outlined above. In Barras Gug (West) on the north side of the isthmus the faults converge into one main dislocation which is a gouge-filled fissure, quite incapable of resisting erosion if it were more exposed to the prevailing wind direction.

The west face of the buttress, upon which stand the Outer Wards of the Castle, has been largely formed by the slipping away of rock from steeply dipping fault-surfaces. Hanging-wall remnants locally adhere to the fault-planes, and rock-falls occur from time to time. The erosion of the ridge of land connecting The Island with the mainland has been accelerated by the numerous fault-planes which cross the gap. According to Radford (1936, pp. 3–4), when the Castle was first built, in about 1145, parts of it 'stood on the isthmus which at that time connected the headland with the mainland'. In 1235–40 the wards on the mainland were built: 'At this period a bridge connected the two parts of the Castle. ... By the sixteenth century the gap between the island and the mainland had widened to almost its present proportions.' Rock-falls from the gap have happened recently, and on the south side the slope as a whole is very insecure. The writer has been asked by workmen on the Castle whether a through-cave ever existed below the gap. It seems quite possible, but fallen rock and rubble mask the evidence for or against.

Merlin's Cave, however, passes from Tintagel Haven, below the Castle on The Island, to the West Cove. It has been eroded along a normal fault which is a member of the Caves Fault Zone. In this case it is doubtful whether the waves in Tintagel Haven were actually responsible for driving the whole tunnel. The present-day erosion in the cave is almost entirely due to seas piling in from the south-west end, and it seems probable that such waves were dominantly responsible for the cave. Even so, waves in the Haven may quite well have started eroding along the north-east end of the fault-plane, and they probably played a considerable part in enlarging the cave entrance. It is of interest to note that above Merlin's Cave, at its south-west end, there is another smaller cave containing striking slicken-sided grooving and accessible from a ledge. One is tempted to speculate on the age of the sea-level which was responsible for the formation of this little feature.

West Cove owes its origin almost entirely to excavation along the strike of several fault-planes. Of these, the Castle Fault is the most important. This dislocation, rising from beneath the beach deposits, runs up the re-entrant in the east corner of the Cove to the Castle Gateway. The effect of undercutting by the sea at the base of the plane of weakness due to the fault, and also to fractures formed by *ac*-joints striking N.25°–30°W., have resulted in the removal of a great downward-tapering wedge of rock from below the Outer Wards of the Castle. Somewhat similar re-entrants on a smaller scale have been known to form on the sides of artificial excavations which cut obliquely across fault zones (McCallum, 1930; Blyth, 1943, p. 251). The vertical wall, upon which stand the Outer Wards, abuts against a sloping, grass-covered surface, the foot-wall of the Castle Fault. At the foot of this grassy slope is a sea cliff which, at the back of the Cove, is merely a step, but towards the sea rapidly rises into a formidable precipice. This side of the West Cove is therefore a youthful hog's-back which cuts across the older slope-over-wall feature that forms the main cliff-line extending to the south.

The fact that the wall of the Castle gateway defences – which for obvious reasons must have been built along the extreme edge of the vertical precipice – is still standing, was considered by Guilcher (1949, p. 714 and Fig. 10ᴇ) as evidence that the cliffs were becoming stabilised. The Castle wall, however, is only unbroken where the base of the vertical cliff is resting on the inclined fault-plane. Where the precipice descends sheer to the West Cove beach and its base is more exposed to marine attack, the original Castle wall above has disappeared. The low wall which now separates the visitor from *un gouffre vertigineux* has only recently been constructed along an edge which had broken away. West Cove beach is covered with very large boulders, and high tide now barely reaches the foot of the precipice. The debris of ancient rock-falls is still to a great extent protecting the base of the cliff from erosion. Waves will arm themselves with small boulders or rock fragments, but their erosive force is checked by big ones. 'The secret of power . . . is not the big stick. It's the liftable stick' (Kipling).

The tunnelling of Merlin's Cave and the similar formation of a big, more or less parallel, but inaccessible cavern between it and Tintagel Head (see Dewey, 1909, Fig. 5; 1914, pl. 4), suggest the tactics of the wave attack which evolved the West Cove. The waves driving up from the south-west worked their way along the various shatter zones, with the Castle Fault as their south-eastern boundary. There are here several fault zones and the tunnels would not be widely spaced, but would rather tend to connect and collapse into each other. Erosion would also be active along the north-north-westerly *ac*-joints which are strongly developed hereabouts, with the consequent breaking away of the cave portals to form a retreating cliff similar to that now seen below the Outer Wards of the Castle. This modern cliff is itself controlled by these cross-joints. The narrow Cove marks the width of the frontal wave attack; behind, that is to the south, the cliffs depart from the plane of the Castle Fault; they swing southerly towards Lambshouse Quarry. Along this stretch they are unbroken by faults, and their slope-over-wall profile is irregular. The original sloping cliff formed by the stripping of the fault-plane has retreated parallel to itself, probably under subaerial periglacial erosion, and the present-day cliff is the result of 'freshening' by wave attack, as Cotton (1951) suggests, during post-glacial or inter-glacial stages.

There is little doubt that The Island, a salient of the coastline, owes its existence to the fact that the base of its cliffs to the south, west, and north is composed of resistant flat-lying volcanic rocks. On these rest the phyllites, which, as a klippe, form the main mass of the headland. The present-day marine action seems to be having little effect on the tough rocks forming the seaward face of The Island. On the landward side, however, where the cliffs are cut by fault zones, the sedimentary beds below the Volcanic Series are brought up into the range of wave erosion, and caves are being driven into and through the promontory. In time, the south-east side, with its Castle ruins, will eventually collapse, and The Island will fully justify its title.

Southwards from The Island, and for a distance of about three miles, the points of headlands all lie on an approximately straight ine. This line is indented by West Cove in the north, the beach of Trebarwith Strand, and by Backway's Cove in the south, beyond which it can be seen to run close, and parallel, to the cliffs as far as Tregonnic Tail, about one mile south of Dennis Point. This line of headlands, seen in plan, is analogous to the accordance of mountain summits which one sees in profile. It marks the general position of an older shoreline now locally embayed by more youthful retrogression.

The open bay of Trebarwith Strand lies between the headlands of Penhallic Point on the north and Dennis Point on the south (Fig. 6.9). The cliffs are stepped back about 600 yds at Hole Beach and 200 yds at Port William from the line tangential to the two points. The stretch between Penhallic Point and Hole Beach is composed of closely jointed Delabole Slate overlain by Woolgarden Phyllites. It is strongly indented, and the cliff foot is largely hidden by rock-fall talus. At Hole Beach are two converging normal faults which strike south-south-westerly. Their seaward prolongation passes between Dennis Point and Gull Island, and would cross the line joining the headlands opposite the mouth of the Trebarwith Strand valley. In the cliffs south of Hole Beach these faults are responsible for the reappearance of the Volcanic Series overlain by sheared phyllites, above which appears the Flat Thrust (T_1). Above the thrust are Delabole Slates, now extensively quarried, and Woolgarden Phyllites. The volcanic rocks form a sloping wave-eroded apron at the foot of the cliffs, and they, together with the sedimentary rocks above them, dip gently westwards. All these strata are broken by other normal faults at Vean Hole, Lill Cove and in Port William. At this last locality Tredorn Phyllites are thrown down below sea-level and form Dennis Point; the faults responsible can be seen running diagonally up the great 300-ft cliff which flanks the point. An extensive sandy beach now covers the foreshore of the embayment, but at

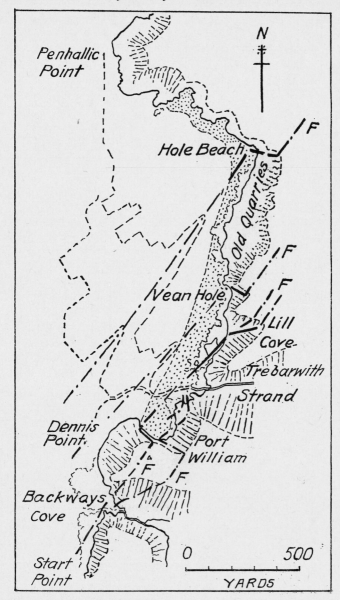

Fig. 6.9 The evolution of the Trebarwith Strand embayment by erosion guided by zones of normal faulting (F)

high tide the waves break locally against the phyllites of the vertical cliff above its plinth of volcanic rocks.

The mouth of the prolonged Trebarwith Strand stream must once have formed some kind of re-entrant in the initial coastline; and, as stated above, this inlet approximately coincided with the point where the Hole Beach faults crossed this coast. On the seaward side of the fault zone the rocks would consist of slates and phyllites; on the landward side there would be similar rock types resting on the sheared phyllites of the Flat Thrust (T_1). Powerful storm waves driven from the south-west – the direction of maximum fetch – would attack directly along the line of the faults and shattered rocks. The attack thus proceeded in the same manner as that recognised at West Cove, and there was a general retrogression from the south-south-west towards the present cliff-line between Penhallic Point and Hole Beach. The main line of marine advance was probably along the zone of dislocation, and would form a gully having an asymmetrical cross-section. The west side would be steep and undercut, the east side inclined in conformity with the fault-planes and, to a certain extent, with the local dip. Rock-falls and landslips, from the two sides respectively, must have kept the waves armed with debris, and steady retreat of the cliffs would occur. In this way the other fault zones were exposed, and the attack also proceeded along them, as can be observed on the Vean Hole and Lill Cove faults at the present time. On each of the fault groups the cliff-line can be seen to be stepped back, with the formation of a re-entrant of which one side is the fault-plane. At Hole Beach the sloping cliff face has retreated back from, and below, the normal fault-plane. Such retreat of the cliff parallel to itself probably accounts for the slope-over-wall form – now partly destroyed by quarrying – which characterised the original cliff between Hole Beach and Trebarwith Strand (Balchin, 1946, p. 334).

The embayment has thus been eroded by attack along oblique zones of weakness with the concomitant removal of the material lying between parallel faults. As long as the attack was on rocks composed of slates and phyllites and their sheared equivalents, erosion must have been fairly easy. But now, from Hole Beach to Port William, the exposed base of the sea cliff is composed of the more resistant Volcanic Series, dipping gently to the west and in which jointing is widely spaced. These rocks are to a considerable extent protecting the coast from rapid retrogression. Nevertheless they too are not immune from erosion, and large blocks detached along joint-planes can be seen to have slipped down the seaward-dipping planes of stratification.

Port William presents an excellent example of wave erosion guided by normal faults which dip about 40° to 50° towards N.75°W. Each fault has a cave eroded at its foot. The outer, most westerly, faults connect with those at Lill Cove, and a gully behind a stack on the shore near the latter marks the trace of the dislocation. The whole hillside forming the back or east face of Port William coincides with the foot-wall surface of the lowest fault. Resting on this exposed fault-plane are blocks of rock, loose rubble and soil which require little encouragement to slip downwards. Wave attack, now cutting into the jointed and down-dragged rocks at the foot of the slope, is actively eroding here at the present time, and the writer has several times found rocks which have freshly fallen between successive visits. The hog's-back cliff which is now forming is in an infantile stage!

The strike of the normal fault system, which has been so important in guiding the wave attack, gradually swings more southerly as the coast is followed towards the south. This is reflected in the form of the coastline itself. At Barras Gug fault-line chasms and coves are oriented at right-angles to the coast. At Trebarwith Strand the re-entrants eroded along fault-planes are at acute angles to the general line of cliffs; but south of Backway's Cove the regular coast as a whole corresponds to the general projected trend of the fault-planes. The Treligga Cliffs are the result of marine attack working its way by slow piecemeal erosion, unassisted by any marked zones of weakness, into the foot-wall of a fault zone which is now largely lost from view. This zone is visible in Backways Cove and, at the south side of the inlet, shows as a single fault face over 100 ft high that has been stripped by rock-falls during the last few years. The influence of such structural planes can also be seen at the south end of the cliffs, where there is extensive landslipping at Tregonnic Tail, and the cave of Flat Hole is eroded along a fault.[1] The Treligga Cliffs thus owe their regular plan to the fact that they are fault-line cliffs which have suffered but slight retrogression. Similar conditions probably persist even further south, where the cliff-line follows a regular, almost straight, course as far as Port Isaac Bay.

4. CONCLUSIONS

The coastal morphology of the Tintagel district of north Cornwall

[1] About 1,500 and 1,200 yds south of Backways Cove respectively.

is largely dependent on the geological structure of the rocks. Balchin (1946) considered that the north Cornish cliffs were, from a regional standpoint, old and mature. Locally, however, they present many youthful features, and the marine erosion, begun so long ago, is now believed by the writer to have followed in principle the tactics of the present-day wave attack on the coast. Bedding planes, joint-planes and normal faults all formed zones of weakness in the rocks which were picked out by the waves, and the crenulations of the present coast are seen to be controlled by such features. Normal faults appear to have been the most important, especially where they trend at an acute angle to the present coast, and are more or less parallel in strike to the direction of maximum fetch. Erosion, having once begun to make inroads along parallel fault zones, has been able to cut back the cliff-line by the collapse and wearing away of the rock slices between the inclined shatter-belts.

It was in this way that the cliffs from The Island to Treligga Common were evolved. The strike of the faults diverges slightly from the line of the cliffs, and the dislocations all dip seawards at angles of about 45°. Under such circumstances the cliff development started with erosion along and the stripping of the fault-plane, and the formation of a sloping surface. Wave action, attacking at the foot of this slope, undercut it and produced typical slope-over-wall or hog's-back cliffs (Fig. 6.10(*a*) and (*b*), which, if the marine activity were powerful, might in turn be destroyed, with the formation of a flat-topped cliff (Fig. 6.10(*c*); Arber, 1949, p. 194). Alternatively, as proposed by Cotton (1951), when the sea-level fell during a period of glaciation, the cliff would be subjected to sub-aerial periglacial erosion. This would result in flat-topped cliffs becoming bevelled, and in original hog's-back cliffs being eroded back or retreating parallel to themselves (Bryan, 1940; Wood, 1942; Penck, summarised in Engeln, 1949, pp. 263–4). In this way the initial slope would be retained, and also extended downwards at the expense of the vertical sea cliff; at the same time the upper edge would tend to become rounded. Eventually the retreating hog's-back would, by erosion above and accumulation below, change to an even continuous slope (Fig. 6.10(*d*)). With the melting of the glacial ice, marine erosion of the cliff base would begin again as the sea-level rose. Head and talus which had accumulated would be washed away, and the vertical sea cliff would be disinterred or developed again by 'freshening' (Fig. 6.10(*e*)). Professor Cotton's proposed theory of two-cycle origin for these cliffs explains many of the features observed in this area.

Short stretches of the coast are still recognisable as being true fault-line cliffs. Elsewhere, where the cliffs swing gently back from the line of dislocation, one may find the same slope-over-wall form, but it is of a less clean-cut appearance. These latter cliffs are believed to be eroded fault-line features which have retreated, but have still retained, in a slightly modified form, their original profiles. Such retreat and modification of hog's-back cliffs explains much of the

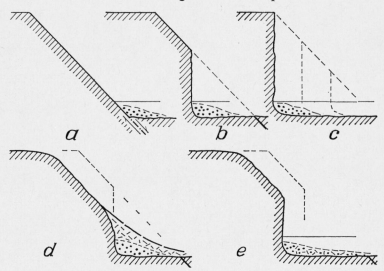

Fig. 6.10 The development and retreat of a fault-line cliff

a to *c* by marine erosion; *d* and *e* by two-cycle process, as suggested by Cotton (1951).

a. – The stripping of an inclined fault-plane.
b. – Wave-erosion develops a slope-over-wall cliff.
c. – Continued wave-action destroys the slope, and a flat-topped cliff is formed.
d. – Sub-aerial weathering during a period of lowered sea-level results in the retreat of the cliff, and the accumulation of head on top of marine deposits.
e. – Return of the sea rejuvenates or 'freshens' the lower face of the cliff.

evolution of the present coastline and accounts for the *raison d'être* of many of the features encountered south of Tintagel.

North and east of The Island the coast changes in character, and in the same area the structures begin to swing around the nose of the Davidstow Anticline. The coast between The Island and Bossiney Haven is crenulated with relatively large havens and headlands, and narrow steeply-sided inlets. The wave attack is here eroding isthmuses along normal faults and soft beds. It is also penetrating inland along the joint-planes which traverse the rocks in various directions.

The effect of erosion guided by jointing, almost to the exclusion of other structures, becomes more and more evident as one proceeds northwards from Bossiney Haven. Sloping cliffs, which resulted from the stripping of strong inclined joints, stacks isolated by attack along vertical joints and the forms of bays, headlands and blow-holes, all show the effect of the same type of controlling influence.

Over different parts of this coastal stretch the geological structures which guided the wave attack, varied in character and orientation, while the attack itself varied in direction and intensity relative to the shoreline. On a coast such as this, where variety in morphology has made the cliff scenery famous for centuries, no generalisation can be applied to the local evolution of the cliffs. The development of form of each headland, bay or line of cliff is guided by the geological structures and by their relationship to the direction of effective wave attack. The rule of *suum cuique* seems to be as applicable here as it is in other branches of geology.

REFERENCES

ARBER, E. A. N. (1911) *The Coast Scenery of North Devon.*

ARBER, M. A. (1949) 'Cliff profiles of Devon and Cornwall', *Geogr. J.*, CXIV 191.

BALCHIN, W. G. V. (1937) 'The erosion surfaces of North Cornwall', *Geogr. J.*, XC 52.

—— (1946) 'The geomorphology of the North Cornish coast', *Trans. Roy. Geol. Soc. Cornwall*, XVII 317.

BLYTH, F. G. H. (1943) *A Geology for Engineers*, 1st ed., p. 251.

BRYAN, K. (1940) 'The retreat of slopes', *Am. Assoc. Amer. Geogr.*, XXX 254.

CHALLINOR, J. (1931) 'Some coastal features of North Cardiganshire', *Geol. Mag.*, LXVIII 111.

CLOOS, E. (1937) 'The application of recent structural methods in the crystalline rocks of Maryland', *Md. Geol. Surv.*, XIII (1) 27.

COTTON, C. A. (1951) 'Atlantic gulfs, estuaries and cliffs', *Geol. Mag.*, LXXXVIII 113.

DE LA BECHE, H. (1839) 'Report on the geology of Cornwall, Devon and West Somerset', *Geol. Surv.*

DEWEY, H. (1909) 'On overthrusts at Tintagel (North Cornwall)', *Quart. J. Geol. Soc. Lond.*, LXV 265.

—— (1914) 'The geology of North Cornwall', *Proc. Geol. Assoc.*, XXV 154.

—— (1916) 'On the origin of some river-gorges in Cornwall and Devon', *Quart. J. Geol. Soc. Lond.*, LXXII 63.

—— (1935) 'South-west England: British Regional Geology', *Geol. Surv.*

ENGELN, O. D. VON (1949) *Geomorphology* (New York).

FOURMARIER, P. (1936) 'Observations préliminaires sur le clivage schisteux entre Boscastle et Newquay (Cornwall, Angleterre)', *Bull. Acad. Roy. Belg. Cl. Sci.*, 5ᵉ sér., XXII 1026.

GUILCHER, A. (1949) 'Aspects et problèmes morphologiques du Massif de Devon-Cornwall comparés à ceux d'Armorique', *Rev. Géogr. Alpine*, XXXVII 689.

—— (1950) 'Nivation, cryoplantion et solifluxion quaternaires dans les collines de Bretagne occidentale et du Nord du Devonshire', *Rev. Géomorph. dynamique*, I, no. 2, 53.

MCCALLUM, R. T. (1930–1) 'The opening-out of Cofton Tunnel, London, Midland and Scottish Railway', *Proc. Inst. C.E.*, CCXXXI 161.

OWEN, D. E. (1934) 'The Carboniferous rocks of the North Cornish coast and their structures', *Proc. Geol. Assoc.*, XLV 451.

—— (1950) 'Carboniferous deposits in Cornubia', *Trans. Roy. Geol. Soc., Cornwall*, XVIII (1949) 65.

PARKINSON, J. (1903) 'The geology of the Tintagel and Davidstow district (northern Cornwall)', *Quart. J. Geol. Soc. Lond.*, LIX 408.

PATTISON, S. R. (1847) 'On the geology of the Tintagel district', *Thirty-fourth Ann. Rept Roy. Geol. Soc. Cornwall*, 3.

RADFORD, C. A. R. (1936) 'Tintagel Castle, Cornwall', *H.M. Office of Works* (H.M.S.O., London).

WHITAKER, W. (1911) in *Royal Commission on Coast Erosion and Afforestation* (H.M.S.O., London) III (1) 6.

WILLIS, B., and WILLIS, R. (1934) *Geologic Structures*, 3rd ed. (New York).

WILSON, G. (1946) 'The relationship of slaty cleavage and kindred structures to tectonics', *Proc. Geol. Assoc.*, LVII 263.

—— (1951) 'The tectonics of the Tintagel area, North Cornwall', *Quart. J. Geol. Soc. Lond.*, CVI 393.

WOOD, A. (1942) 'The development of hillside slopes', *Proc. Geol. Assoc.*, LIII 128.

ICD F

7 The Form of Nantasket Beach

DOUGLAS W. JOHNSON AND
WILLIAM G. REED, JR

INTRODUCTION

THIS paper presents the results of a study of the form of Nantasket
Beach, and includes a discussion of the stages of development through
which the beach has passed to reach its present form, and of the
processes by which that development has been accomplished. Our
attention was directed to the Nantasket problem by Professor W. M.
Davis, who was the first to discover the significance of the abandoned
marine cliffs and beaches, and their relation to islands which have
long since disappeared. Acknowledgements are also due to the Boston
street commissioners for assistance in securing old maps for examination; to Mr E. G. Knight of Hull for information regarding conditions prior to the building of the County Road on Nantasket
Beach; and to Mr F. M. Hersey of Boston, and the officials of the
United States Coast and Geodetic Survey, for numerous courtesies.
In addition to these gentlemen, our thanks are due to many others
in Boston and Hull for various services.

Nantasket Beach lies at the south-eastern border of Boston
Harbour, separating that portion of the harbour from the Atlantic
Ocean (Fig. 7.1). The name 'Nantasket Beach' is generally applied
to all of that irregularly shaped lowland between the rocky hill of
the Atlantic on the south, and the drumlin known as Allerton Great
Hill on the north, and is not restricted to that portion of the lowland
immediately bordering the ocean at the present time. As thus
defined, Nantasket Beach has a width of from a few hundred feet to
more than half a mile, and a length of a little more than three miles.
In our discussion we include the neighbouring district of Hull, as
well as several outlying islands, which are more or less closely
related to certain phases of our investigation.

THE PROBLEM STATED

A casual study of the Nantasket district makes clear the fact that
Nantasket Beach consists of sand, gravel and cobbles, deposited by

wave action between several drumlins which formerly existed as islands. The problem which we have to consider may therefore be described as a problem in island-tying by means of beaches. The tying of islands to each other and to the mainland, by the formation of connecting beaches, has been recognised as a common phenomenon along a youthful shoreline of depression, where islands are

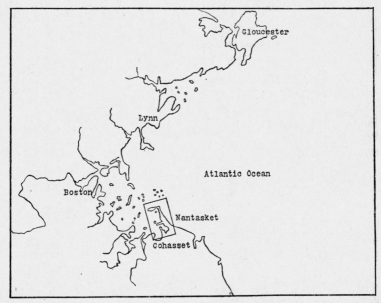

Fig. 7.1   Location of Nantasket area

apt to be more or less numerous. Boston Harbour occurs on a shoreline of depression, but the islands which help to form the harbour, and which are frequently connected with each other and with the mainland by beaches, do not as a rule represent the summits of hills left as islands by the depression of a maturely dissected mainland. They are for the most part typical drumlins, the trend of whose long axes indicates that the ice sheet which fashioned them moved from the land south-eastward out to sea. It is evident that drumlin islands might be formed along a shoreline of elevation; hence the phenomena about to be described might occur along both of the standard types of shorelines. The principles involved in our discussion remain the same, whether the islands be composed of solid rock or unconsolidated glacial till; but it will appear that the stages of shoreline development are passed through more quickly, the wave-cut cliffs are more symmetrical and the past conditions more easily reconstructed where drumlin islands are involved, as in the Nantasket case.

In the following pages we shall briefly review the principles of shoreline development, and then describe in some detail the present form of Nantasket Beach. On the basis of this description, and in view of the principles of shoreline development, we shall endeavour to reconstruct the initial form of the Nantasket district. Still guided by the principles of shoreline development, we shall next trace the successive steps in the development of Nantasket Beach from the initial form to the present form, with brief attention to the changes which will probably occur in the future. It will appear that Nantasket Beach is a very complicated example of island-tying, which illustrates in a remarkable manner the fact that shorelines are the product of systematic evolution according to definite physiographic laws.

LITERATURE

So far as we are aware, no detailed account of the physiography of Nantasket has been published. Professor W. O. Crosby (1893) has described the hard rock geology of the district just south of the beach in great detail and has considered the beach and drumlins at some length. He has also discussed the evidence of post-glacial changes of level in the Nantasket area. Other references to the district here described are found throughout the literature on the Boston Basin, but are not of importance in the present discussion.

In 1896 Professor W. M. Davis published a paper entitled 'The Outline of Cape Cod', in which he discussed at some length the principles of wave and current action, and applied these principles in a study of the present form of Cape Cod and the past changes in the outline of the cape. The principles set forth in Professor Davis's essay are considered more fully on a later page.

Dr F. P. Gulliver, in a paper on 'Shoreline Topography' (1899), has discussed at length various shore forms, including beaches which connect islands with the mainland or with each other. To such beaches he has applied the name 'tombolo'. Several types of tombolos are described, and Nantasket Beach, which might be described as a complex tombolo, is briefly mentioned.

In common with all students of shoreline topography, we are indebted to Dr G. K. Gilbert's classic studies of lake shores for the elucidation of many of the principles upon which all shoreline studies must be based.

A brief note on 'The Geology of the Nantasket Area', containing an outline of the physiography of the district, was published by Professor D. W. Johnson in *Science* three years ago.

THE PRINCIPLES OF SHORELINE DEVELOPMENT

Physiographers recognise two distinct classes of shorelines – those formed by a (relative) elevation of the land, called shorelines of elevation; and those formed by a (relative) depression of the land, called shorelines of depression. It is not necessary to repeat the characteristic features of these two classes of shorelines, nor to trace the successive stages by which young shorelines of each class acquire, by the time they reach maturity, curves of a relatively simple pattern, marine cliffs more or less bold, and shelving beaches at the foot of the cliffs. Initial characteristics and stages of development are both set forth in our best textbooks on physiography.

We may note, however, that the processes of shoreline development involve both wave and current action. It has been shown that wave action is largely confined to the erosion of the land margins, to the transportation of the eroded material a short distance from the shoreline and to the deposition of the material in the deeper water; and to the heaping-up of sand, gravel and cobbles into long ridges or beaches, where the conditions favour wave building more than wave erosion. Current action, on the other hand, effects but little erosion, and is mainly effective in the longshore transportation of material previously eroded by the waves or brought in by rivers. It has been shown that the combined effects of these two processes is to produce, in time, a shoreline characterised by long, simple curves, and free from sharp angles or other irregularities. Headlands are cut back, or retrograded, and re-entrants are built forward, or prograded, in the attempt made by waves and currents to straighten out the initial irregularities of the shores, and thus to establish a graded shoreline. The process is analogous to the formation of graded stream profiles by the degrading of elevations and the aggrading of depressions. If the waves cut back faster on one side of a headland than they do at the headland or beyond, and a strong current sweeps along the shore, the formation of a sharp angle at the headland is prevented by the prograding of the shore beyond the headland as rapidly as it is retrograded in the region of pronounced cutting. Thus, in the case of Cape Cod, Professor Davis has shown that the retrograding of the shore in the vicinity of Highland Light, due to active wave erosion, has been accompanied by a prograding of the shore father north, where successive beaches have been built forwards to maintain the simple curvature of a maturely graded shoreline. Equilibrium is reached, or the beach is maturely graded, when a gently curved or straight shoreline is developed; thereafter the headlands and beaches both retrograde gradually under the continued attack of the waves.

Inasmuch as the Nantasket problem involves a complex example of island-tying, we may here consider certain principles underlying the formation of connecting beaches, or tying bars as they are often called. If an island faces a large expanse of open water on which large waves are produced, and these waves come in general from one direction, the end of the island exposed to the brunt of the wave attack will be eroded, and the eroded material will be gradually drifted back along the sides of the island and strung out behind as a spit. In course of time the spit may reach the mainland or another island, and the island-tying is complete. Variations in local conditions may result in various forms of the tying bar; examples of several forms are described by Dr Gulliver. It is possible that in some cases the bar may be built from the mainland out to the island (Gulliver, 1899, p. 192).

Backward tying is not the only form of island-tying to be observed along the shores. Lateral tying is certainly strongly developed in the Boston region, and we believe that many cases now regarded as examples of simple backward tying will prove to be more or less complicated examples of lateral tying. If wave erosion is most active on the eastern end of an island which lies at the mouth of a bay, and which is between two headlands situated to the north and south of the island (Fig. 7.2), several types of tying may result under different conditions. If there is no pronounced current action except the onshore and offshore tidal currents and the movements of the water due to waves coming from the east, simple backward tying may result (a). Under the same conditions as those just outlined, provided that the shallowing of the water landward is favourable, the material eroded from the head of the island may be drifted backwards but at the same time northwards and southwards in curving lines along a zone where the water is of such depth as to favour deposition before reaching the bay-head. Similar deposits from the headlands would meet those from the island, and lateral tying by curved bars would result (*b*). These bars might then be prograded, as explained on a previous page, until they formed a nearly straight shoreline between the island and the headlands. If the tidal action were fairly strong, the bars from the island and headlands might not join, leaving each portion as a spit, possibly more or less irregular in form, at its free end (*c*). If a pronounced longshore current existed, the waves still

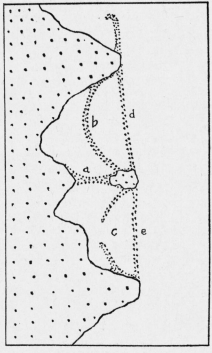

Fig. 7.2

coming from the east, the material eroded from the island by wave action might be transported by the current from the island towards the headland, building a bar which would eventually tie the island to the mainland (*d*). In a similar manner a bar built northwards from the southern headland might effect the tying of the island to the mainland (*e*). A sufficiently strong tide might prevent the tying in either case by maintaining a tidal inlet. But if the tying were effected, it should be regarded as lateral tying, the connecting bar being built at right-angles to the direction of wave attack, and parallel to the shoreline. Lateral tying similar to that represented by (*d*) and (*e*) in

Fig. 7.2 might be produced if the character of the sea bottom caused the waves to break along the line *de* instead of entering the bay to the points *b* and *c*; also in case a barrier beach migrating towards the shore encountered the island and headlands in its progress.

If several islands instead of one were involved in the foregoing cases, more complex types of backward and lateral tying would result. Nantasket Beach represents a complicated case of both lateral and backward tying, involving for the most part prograded lateral tying bars of the types *b* and *c* of Fig. 7.2.

THE PRESENT FORM OF NANTASKET BEACH

The principal topographic features of Nantasket Beach are shown on Fig. 7.7., based on a chart of Boston Harbour prepared by the United States Coast and Geodetic Survey (No. 246, C. & G. S. Boston Harbour, 1907). The larger features appear on the chart, but we have added the details of smaller beaches, wave-cut cliffs, etc. The irregular hills in the southern part of the map are composed of much altered sedimentary and igneous rocks which are very resistant and yield but slowly to the attack of the waves and weather· All other elevations on the map represented by contours are drumlins more or less eroded by wave action. The lower areas, including the lower ridges indicated by short hachures, are practically all of beach material; the exceptions consist of low areas of till between certain drumlins located close to each other, beach sand gathered into small dunes by the wind, and some deposits in swampy areas to be considered later. If we except the rock above-mentioned, we may properly say that the features of the Nantasket region are due to marine action upon drumlins; for the effects of stream action and wind action are so slight as to be negligible.

The drumlins

In describing the present form of the Nantasket drumlins it will be convenient to consider them in the order of their preservation from marine erosion. The letters in parentheses refer to the respective drumlins on the map (Fig. 7.7.) The best preserved of the Nantasket drumlins is a small one called Hampton Hill (H), located in the southern part of the region, back of the beach. It has been slightly cliffed by the harbour waves, on the south-west, but is otherwise practically in the same condition as when the ice left it. Nantasket

Hill (N) at Hull, also called Telegraph Hill, is another drumlin which has suffered but little erosion; it is slightly cliffed on the south. Thornbush Hill (T), just west of Nantasket Hill, is somewhat more strongly cliffed, but retains its initial form to a marked degree. The erosion has taken place at the south-west side. Sagamore Head (Sa), near Hampton Hill, preserves a nearly perfect outline except for a pronounced cliff on the north-east side and a minor cliff on the north and west. The main cliff is well back from the present shoreline, and has evidently not been touched by the waves for many years.

North of Sagamore Head is White Head (W), a drumlin which retains its initial form fairly well on the south, although a slight cliffing is noticeable there, but which has a remarkable strongly curved cliff cut into its northern side, and smaller cliffs on the north-east and east. Like the north-east cliff on Sagamore Head, the cliffs on the north and east sides of White Head are well back from the present shoreline and have long remained untouched by the waves. West of White Head are several low drumloidal hills, connected by lower areas of till and cliffed on both the north and south sides. Great Hill (G), at Allerton, has a strongly marked cliff on the eastern end where the waves are still cutting into the hill, although not so effectively as formerly. There has apparently been a slight cliffing on the western end of Great Hill also. Strawberry Hill (St), about half-way between Allerton and Sagamore Head, is in many respects the most remarkable drumlin in the district. Except for a short distance along the north-west side, it has been cliffed throughout its entire circumference; a rather inconspicuous cliff is developed along the north side, more prominent cliffs on the south and west sides, while the south-east face is a splendid marine cliff long ago abandoned by the waves. In fact the only point where the sea still reaches the drumlin is along its south-west side. There is a marked escarpment on the north-east corner of the cliffed drumlin, but much of this is due to the removal of till for road-building. Professor Isaiah Bowman informs us, however, that a small nip existed there before the excavations by man obscured the relations. It should be noted that the abandoned cliffs of Sagamore Head, White Head and Strawberry Hill do not face in the direction of the present shoreline, but make pronounced angles with that shoreline, as shown by the map.

Quarter Ledge (Q) at Hull is a more than half-consumed drumlin, the marine cliff facing northwards. Little Hill (L) at Allerton is of

ICD F 2

special interest because it is evidently but a small remnant of a drumlin on the north-east of Great Hill. It would doubtless have been completely removed by the waves ere this but for the protection afforded by a stone sea-wall constructed north and east of it to prevent its complete destruction. Skull Head (Sk) represents the final stage in the series, having been completely destroyed. This drumlin was situated to the north-west of Strawberry Hill, and so far as we can tell was probably of small size. It was apparently nearly destroyed by wave action from the west, the last remnant being removed by man and used as road material. Those who remember this drumlin remnant agree in describing it as having a gentle slope toward the east and a steep cliff facing west. The presence of great boulders near the supposed former site of this drumlin, the shape of the associated beaches and the westward protuberance of the shoreline north-west of Strawberry Hill confirm the descriptions and location of the drumlin given by the inhabitants.

West of Nantasket Beach there are many drumlins more or less cliffed by marine erosion. On Nantasket Beach are drumlins in all stages of marine erosion, from slight cliffing to almost complete destruction. East of Nantasket Beach no drumlins are encountered. The suggestion is very strong that the sudden cessation of drumlins to the east is due to the complete removal of formerly existing drumlins by marine erosion. As will appear later, there is strong evidence in favour of this interpretation.

The beaches

Under this head are described the various spits, connecting bars, beaches, etc., both ancient and recent, which make up the composite feature called Nantasket Beach.

The beaches at Hull present no striking characteristics. The cliffed drumlins of Nantasket Hill, Thornbush Hill and Quarter Ledge are close together, connected by lowland areas of till, and the cliffed portions are bordered by a narrow, sometimes bouldery beach. A sand spit, called Windmill Point (WP), is strung out towards the west, probably under the influence of tidal currents passing through Nantasket Roads. This group is connected with Allerton by a bar believed to be the result of simple backward tying from Great Hill and Little Hill. The appearance of a Y-bar is due to a railroad embankment built across the end of the bay back of Great Hill in order that the track would not have to be placed in the very exposed position on the seaward side of the Allerton drumlins. The protuberance of beach material from the north-west side of Great Hill is explained later.

From Allerton Great Hill on the north-west to the rock hills of Cohasset on the south-east a relatively straight beach borders the present shoreline. Back of this modern beach one observes parallel ridges of sand, gravel and cobbles, in all respects similar to the higher part of the present beach which is still being acted upon by the waves. Still farther back the ridges become less prominent, until in the central areas of the Nantasket lowland they are scarcely perceptible. Moreover they are no longer parallel to the modern beach, but are strongly curved, concave towards the east. At the extreme west, however, these curved beaches become prominent features once more and are as high in places as the modern beach.

If we examine the older beaches more carefully, we note several significant points. Just south of Allerton Great Hill the high and prominent westernmost beach, which we may call West Beach, is intersected by the modern beach. West Beach does not touch Great Hill and from the curvature of the beach it seems hardly probable that it connected with the former seaward extension of Great Hill. At its southern end West Beach ties to the north-west side of Strawberry Hill, just in front of the only part of the hill which has no bordering marine cliff. From the western side of the beach projects the protuberance of the Skull Head area, which destroys the otherwise symmetrical curve given to this portion of the harbour shoreline. Of the beaches which intervene between West Beach and the modern beach, a few connect with Strawberry Hill, others curve eastward as if to connect with something formerly situated in front of Strawberry Hill, and still others pass in front of the hill to connect with White Head or Sagamore Head farther south; while at the north all converge towards the intersection of West Beach with the modern beach, merging with the former or being cut off by the latter. The waves from the harbour are now attacking West Beach north of Skull Head, giving it a steeper western face, cutting off part of the western convexity and building a small subsidiary beach towards the north. This attack of the harbour waves upon a beach formerly constructed by the powerful Atlantic waves has become so effective that sea-walls have been built in places to prevent further destruction of the old beach.

South of Strawberry Hill the relations are much the same, except that the beaches are less distinct and less regular in outline. The equivalent of West Beach does not connect directly with Strawberry

Hill, but is truncated by a more recent beach or spit which extends southwards from the south-west end of the great cliff on Strawberry Hill. The older main beach curves rather strongly south-west, continues south and south-east in much broken and complicated ridges, and finally spreads out in a broad, indefinite plain of beach material near the western end of White Head. The most prominent beach in this vicinity is one which extends from the eastern point of Strawberry Hill to the eastern end of White Head, and on which the County Road is located for much of the distance between the two hills. Both east and west of the County Road Beach are some fairly well-marked beaches, more or less obscured by sand dunes, especially towards the east. Two of the older beaches between White Head and Sagamore Head are especially prominent and are practically straight. Both north and south of Strawberry Hill the beaches in the central areas, midway between the drumlin hills, are often so low and indistinct as to be nearly or quite imperceptible. In places the detection of the beaches is made easier by a difference in the grass and other vegetation growing on the beach ridges and in the intervening depressions.

THE INITIAL FORM

In our attempt to reconstruct the initial form of Nantasket Beach, we have appealed to three sources for information: (1) some of the older inhabitants who recall the appearance of the beach in earlier days; (2) old maps and charts of the region; (3) the principles of shoreline development applied to the interpretation of the present forms.

Shoreline changes take place with comparative rapidity, and in some cases a man may live to see profound alterations in the outline of the coast on which he lives. Some residents of Nantasket speak of a time when the sea used to come in to the present location of the County Road. It must be remembered, however, that people are apt to be impressed by the unusual, and that some long-past transgression of exceptional storm waves far across the present beach may be responsible for the impression that the sea is now farther removed from the road than it was fifty years ago. As late as 1898, during the 'Portland Storm', breakers crossed the railroad track, which is well back from the present beach. In regard to the former location and general appearance of the remnant of Skull Head drumlin, now completely lost, the descriptions of the older inhabitants agree fairly well, and are corroborated by the physiographic evidence.

The old maps and charts of the region afford some evidence as to the general outline of the beach in earlier years, but prove to be too inaccurate to justify any conclusions as to recent changes in outline. A chart prepared by the United States Coast and Geodetic Survey in 1846 differs in minor points from the more recent charts of the same area, and a comparison of the two might be expected to show changes in the shoreline since 1846. Indeed, such a comparison has been made in connection with a study of cliff retreat at Allerton Great Hill and an estimated retreat of about 2 ft a year has been inferred on the basis of the comparison. A careful study of the two charts in the light of the geological features of the region makes clear the fact that one or both of them are too inaccurate to warrant any conclusions as to changes in shoreline based on such evidence. For example, it appears from the charts that the shoreline along the south-eastern corner of Allerton Great Hill is farther east today than it was in 1846. Now the shoreline at this point is formed by the cliffed face of the hill, and since this hill is a drumlin which could not have been built forward since the glacial epoch, the charts are manifestly not sufficiently accurate to be used in determining recent changes in shoreline. On the other hand, it should be noted that the chart of 1846 indicates a shoreline so nearly like the present shoreline as to warrant the conclusion that the sea has not been materially closer to the County Road in the last sixty years than it is today, except during unusual storms. Indeed, a chart of Boston Harbour published in the fourth part of *The English Pilot* in 1709, while not accurate in details, seems to show that no pronounced changes in the shoreline of Nantasket Beach have occurred in the last two hundred years.

The application of the principles of shoreline development to the interpretation of the present form of Nantasket Beach offers the only means of determining the initial form of the beach. We believe that by this means it is possible to determine with a fair degree of certainty the geography of the Nantasket region before the present beach came into existence. The problem involves the restoration of the lost drumlins of this portion of Boston Harbour.

There is little difficulty in the restoration of those drumlins which retain their initial form to a considerable degree. The existing drumlins of the Boston district are of the same general type, none of them resembling the greatly elongated type found in some parts of New York. It is possible, therefore, to complete the outlines of Thornbush Hill (T) and Nantasket Hill (N) at Hull, Great Hill (G), Strawberry Hill (St), White Head (W), Sagamore Head (Sa), and Hampton Hill

(H) without danger of appreciable error. This has been done in Fig. 7.3, the restored portions being indicated by broken lines. Where more than half of a drumlin has been destroyed, the restoration cannot be made with the same degree of certainty, and we recognise that the location and size of such drumlins cannot be determined

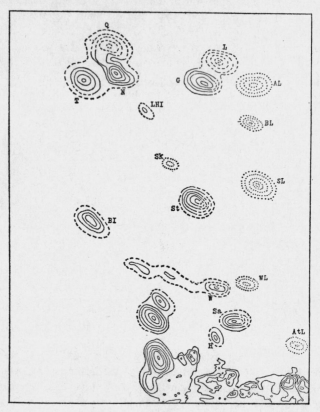

Fig. 7.3

with absolute precision. The margin of error is not so large as materially to affect our problem, and the restorations of Quarter Ledge drumlin (Q) and Allerton Little Hill (L) in Fig. 7.3 (restored portions in broken lines) are believed to be essentially correct. The position of Little Hill will account for the peculiar protuberance of beach material north-west of Great Hill, if we agree that a spit trailing back from Little Hill by the action of waves and currents through Nantasket Roads would have a form somewhat similar to that of Windmill Point (WP) in Fig. 7.7. The restoration of the drumlins which are wholly destroyed involves a larger chance of error, and each individual restoration of this kind must be carefully considered.

The first restoration of a drumlin now completely destroyed (complete restorations shown by dotted lines) is that of Allerton Lost Drumlin (AL). That this drumlin formerly existed is shown by the relations of West Beach. The beach does not connect with Great Hill at the present time, but is abruptly cut off by the present shore-line a short distance south of Great Hill. That this beach formerly continued toward the east seems clear. It is equally clear that the seaward continuation of the beach would not connect with the seaward continuation of Great Hill, unless we imagine the beach to have been bent sharply northwards. This last assumption is contrary to what we should expect in a beach as well developed as West Beach has no evidence to support it, and is one which we are not permitted to make arbitrarily. The precise location of the drumlin with which West Beach must have connected cannot be determined with certainty, nor can its size be inferred; but that it occupied some such position as is indicated in Fig. 7.3 there would seem to be little doubt. It is not permissible to consider West Beach connected with the eastward extension of Little Hill, for this would require a marked northward bend in the beach, or the reconstruction of Little Hill on too large a scale.

It will be convenient to consider the restoration of Skull Head drumlin (Sk) next, as certain features connected with it will aid us in other reconstructions. The location of this drumlin is made clear, as already noted by the peculiar protuberance back of West Beach north-west of Strawberry Hill, by the occurrence of large boulders along the shoreline at this point, and by the historical evidence. That the drumlin was small is indicated by the fact that it has been completely removed, although in a relatively sheltered position, and by the further evidence that this removal was accomplished mainly by the harbour waves, it being stated by those who remember the drumlin that the eastern slope was not cliffed, while the western face was a distinct marine cliff. The last remnant of this drumlin was removed by man in recent years. The location and size of Skull Head drumlin are believed to be essentially correct.

An examination of the great south-east cliff on Strawberry Hill shows that the cliff was formed by waves coming from the south-east, and not from the north-east, the direction of the present wave attack. The fact that a sharp angle on the cliffed drumlin projects forward on that part of the hill which would be most exposed to the direct attack of the waves had no other drumlin existed in front of it to protect it, confirms the opinion that the restoration of a drumlin must be made in the vicinity of the shallow area offshore known as Strawberry Ledge. This we have called the Strawberry Lost Drumlin (SL). As will appear later, the former presence of a drumlin at this point accounts for the north-eastern angle (recently blunted by excavations for road material) of Strawberry Hill, the small amount of cliffing on the north side of the hill, the eastward curve of the beaches north-east of the hill, the direction of the splendid south-east cliff and a certain feature of West Beach to be considered in the next paragraph. Whether the shallow area at Strawberry Ledge has any connection with the Strawberry Lost Drumlin we are unable to say, but that the drumlin must have been located near this spot seems clear.

As has already been noted, Skull Head drumlin (Sk) was apparently not cliffed on the east, or was so slightly cliffed as not to attract the attention of persons who did notice the cliffing on the west. Yet this drumlin must have occupied a position fairly well exposed to the waves of the Atlantic, unless some protection from those waves was afforded by drumlins or beaches farther east. It should be noted also that West Beach is unusually high and broad, the modern beach at the east alone showing the same strength of development. In order that the waves should build a beach so extensive and so well developed, they must either have acted on the ancient West Beach shoreline for a long period of time, or must have been rapidly supplied with an immense amount of material previously reduced to a condition ready for beach construction. That the waves did not act for a long period of time in the vicinity of the ancient West Beach shoreline is shown by the absence of any considerable cliffing on the east end of Skull Head drumlin and the north side of Strawberry Hill. It is evident, moreover, that the large amount of material in West Beach could not have been supplied by the cliffed portions of the existing drumlins in that vicinity, so it must have come from drumlins long ago destroyed, or from the sea bottom. We believe that the most probable condition which will account for all the facts is the former existence of a beach or series of spits more or less completely closing the space of open water between Strawberry Lost Drumlin and Allerton Lost Drumlin, thus forming a barrier which protected Skull Head drumlin and Strawberry Hill from wave action. The construction of this barrier was probably facilitated by the existence of another drumlin in the vicinity of the shallow area east of Bayside, and we have called the restoration of this drumlin (Fig. 7.3) the Bayside Lost Drumlin (BL). As will appear in the next section, the present relation of beaches and cliffs strongly suggests that a drumlin located in the vicinity of the Bayside shallow maintained the barrier so long as any part of the drumlin remained; but that with the complete removal of the drumlin the barrier was broken through, the accumulated debris swept rapidly back to the present position of West Beach, still protecting the east end of Skull Head drumlin but exposing a large part of the north side of Strawberry Hill to the waves which formed the low cliff we observe today.

The highly peculiar character of the cliffing on the north and north-east sides of White Head drumlin can be explained only by the restoration of a drumlin north-east of White Head. This we have called the White Head Lost Drumlin (WL). Its precise location cannot be determined, but it must have been close enough to White Head to control the marked curvatuve of the White Head cliff and the less marked but distinctly curved cliff on the north-east side of Sagamore Head. The position assigned to it in Fig. 7.3 cannot be far from correct. The character of the Sagamore Head cliff, just referred to, necessitates the restoration of another drumlin to the south-east. Unless this drumlin existed somewhere in that region, affording protection to the south-east corner of Sagamore Head, it is difficult to understand why the latter was not cliffed directly from the east, and why the cliff is concave instead of convex. A shallow area north of Atlantic Head may have been the location of this drumlin, as shown in Fig. 7.3. It is possible, however, that it may have been nearer Sagamore Head. We have called this restoration the Atlantic Lost Drumlin (At L).

This completes the restorations which seem required by the present forms of cliffs and beaches. That other drumlins may have existed in the region is, of course, possible; although the former existence of many more in the immediate vicinity of the Nantasket area would doubtless be indicated by peculiar alignments of cliffs on the remaining drumlins, or by the relations of the beaches. That additional drumlins may have existed still farther east is quite possible, but the data necessary for the reconstruction of such easternmost drumlins would be recorded only on drumlins and in beaches since completely destroyed. So far as the present problem is concerned the conditions shown in Fig. 7.3 may fairly be taken to represent the initial one of a series of developmental stages which we will now endeavour to follow until the present form of Nantasket Beach is reached.

THE DEVELOPMENT OF NANTASKET BEACH

In Fig. 7.4 we have endeavoured to represent the conditions which probably existed in the Nantasket region at a much later stage than

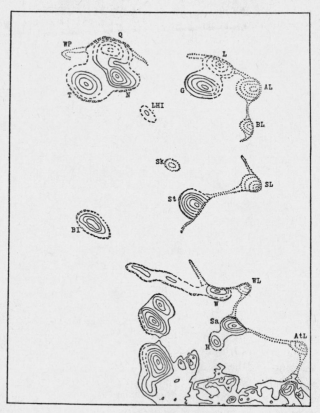

Fig. 7.4.

Fig. 7.3. The Allerton, Bayside and Strawberry Lost drumlins have been much eroded by the waves and the material removed from them has been built into spits or connecting bars, which together with the remaining portions of the drumlins form a barrier to protect the east end of Skull Head drumlin and the north side of Strawberry Hill from any appreciable erosion. From Strawberry Lost Drumlin a bar ties backwards to Strawberry Hill, protecting the north-east corner of the hill and helping to determine the direction of the wave attack which is producing the south-east-facing cliff. That Strawbery Hill and White Head were exposed to strong wave action while Skull Head and Great Hill were well protected is evident from the splendid development of the ancient marine cliff on the two former. White Head Lost Drumlin is much eroded, but still serves to determine the character of the cliff on the north side of White Head, and at the same time effectually to protect the eastern end of the same. Sagamore Head has been cliffed on the north-east, the character of the cliff being determined by the position of White Head Lost Drumlin and Atlantic Lost Drumlin, and the bars tying back from them. Atlantic Lost Drumlin is much eroded, and in addition to being connected with Sagamore Head has a short bar connecting with the rock cliffs just south.

Allerton Little Hill and Quarter Ledge drumlin, facing the main channel to the north, have been considerably eroded, while even the better-protected drumlins have, as a rule, been cliffed slightly, especially on their more exposed sides.

It is evident that some latitude is allowable in the restoration of certain of the features shown in the figure, without affecting the validity of the general interpretation here set forth. For example, the precise shape and location of the sand spits cannot be ascertained; and Atlantic Lost Drumlin might be nearer Sagamore Head, in which case the long bar connecting the two might be altogether absent, or represented by short spits or a short bar. We have indicated, however, those conditions which we consider most probable and the main features of the drawing are believed to be essentially correct.

Fig. 7.5 represents a later stage than Fig. 7.4. The complete destruction of Bayside Lost Drumlin has allowed the material formerly accumulated in its vicinity to be swept back to the Strawberry Hill–Skull Head region, and to be rapidly constructed into the prominent West Beach. At the north this beach still connects with the remaining portion of Allerton Lost Drumlin, thus accounting for the failure of this beach to touch Allerton Great Hill, a relation which is very distinct at the present time. At the south the connection with Strawberry Hill was far enough west to allow a slight cliffing along much of the north side of the hill. The absence of a pronounced cliff at the north-east corner of Strawberry Hill previous to recent

excavations, and the eastward curve of some of the old beaches
north-east of the hill, indicate that a remnant of Strawberry Lost
Drumlin still survived at the period represented by Fig. 7.5 and even
later, preserving the backward-tying bar until West Beach was
considerably prograded. Between Strawberry Hill and White Head

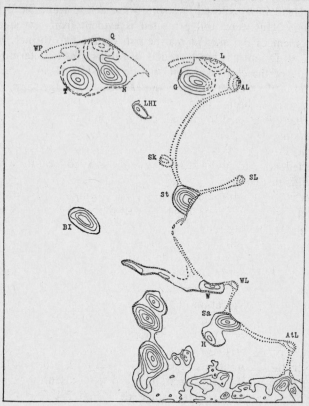

Fig. 7.5

spits or a curved bar nearly or quite close the space of open water,
although the irregular character of the gravel ridges now observable
at this point suggests that the bar may have been repeatedly broken
through during heavy storms. The retreat of the shoreline on the
south-west side of Strawberry Hill has caused the older beach ridge
to be truncated by a sand spit now forming. Other minor developments are indicated, including the continued cliffing of various hills,
and the growth of Windmill Point and other smaller spits.

In the stage represented by Fig. 7.6 the present characteristics of
Nantasket Beach begin to be more easily recognisable. Allerton,
Strawberry, White Head and Atlantic Lost drumlins have all been
completely removed. Prograding has gone on actively in the two
re-entrant curves north and south of Strawberry Hill, the shorelines

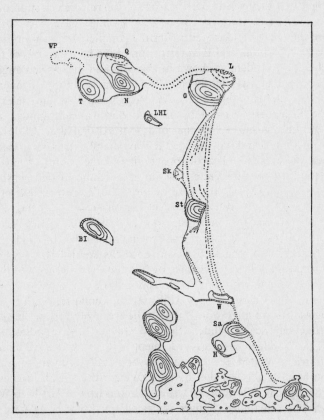

Fig. 7.6.

thus migrating eastwards until a single beach describes a very gently
concave curve from Allerton Little Hill to White Head. That the
process of prograding was relatively rapid is indicated by the small
size of the beaches in the inland areas north and south of Strawberry
Hill. In places these beaches are almost imperceptible, and south of
the hill it seems probable that the change in the position of the
eastern shoreline from the westernmost beach to the County Road

Beach was made without the formation of complete intermediate beaches. That the prograding had proceeded quite far, in the northern re-entrant at least, before the removal of the bar connecting Strawberry Hill with Strawberry Lost Drumlin, is shown by the development of faint beaches just north of the hill, curving eastward so strongly that they would pass in front of the restored portion of the hill if they were prolonged. These beaches must have been formed before the bar was destroyed. After the complete removal of Strawberry Lost Drumlin and the destruction of the bar, the ends of these beaches were eroded, as shown in Fig. 7.6, and the eastern angle of Strawberry Hill was slightly cliffed by the waves. In this manner the portion of the shoreline which had been prograded with reference to the Strawberry Lost Drumlin and bar was retrograded until brought into harmony with the conditions existing after the destruction of drumlin and bar. Before the waves could seriously affect the corner of Strawberry Hill the prograding of the entire beach (from Allerton Little Hill to White Head) as a single unit carried the shoreline eastwards beyond the base of the hill. The prograding of the beach appears to have been connected with the retrograding of the headlands at Allerton and the removal of Whitehead and Atlantic Lost drumlins. As Allerton Lost Drumlin, Little Hill and Great Hill have been cut back, the beaches to the south have been built forward, the point of no change, or fulcrum, being just south of the east end of Allerton Great Hill. The lack of a complete series of beaches south of Strawberry Hill may be connected with a more sudden westward migration of the southern end of the shoreline upon the disappearance of White Head and Atlantic Lost drumlins, and a consequent sudden eastward movement of the zone of wave building just north of White Head drumlin. As soon as the eastward migration of the beaches allowed the shoreline to clear the hill, the successive beaches appear to be more or less continuous from Allerton to White Head. The remaining changes indicated on the drawing need but little comment. The removal of White Head Lost Drumlin, together with the formation of the County Road Beach, has resulted in the cliffing of White Head on the east and north-east, while the removal of the Atlantic Lost Drumlin has allowed the connecting bars to swing back and form a single bar which unites with the rock cliffs at Atlantic Head.

The next stage in the development of Nantasket Beach is that of the present, represented in Fig. 7.7. The principal change from the preceding stage consists in the prograding of the beach until it makes an unbroken, gently curved shoreline from Great Hill to Atlantic Head; the further cliffing of White Head at the eastern end and the abandoning of the cliffs on White Head and Sagamore Head by the waves as the shoreline migrated eastward; the complete removal of the Skull Head drumlin, partly within recent years; the filling-in of

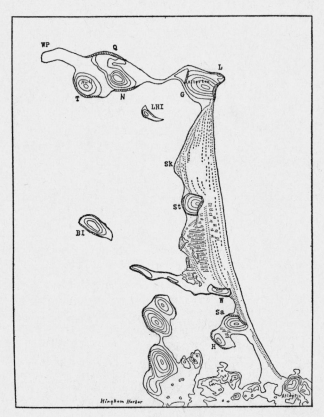

Fig. 7.7

the small bay on the south side of Windmill Point, largely within historic times; and further erosion of all the drumlins still exposed to wave action.

In Fig. 7.8 we have attempted to represent a possible future stage in the development of Nantasket Beach. At the present time the most effective wave erosion is concentrated upon Great Hill and the small remnant of Little Hill. But these hills control the future of the beach,

the erosion of the rocky mainland at the southern end being so slow
as to be practically negligible. Heretofore the retrograding of these
hills has caused the prograding of the beach; at the present time,
however, a condition of equilibrium prevails, and a further cutting-
back of the hills must result in a cutting of the beach also. With

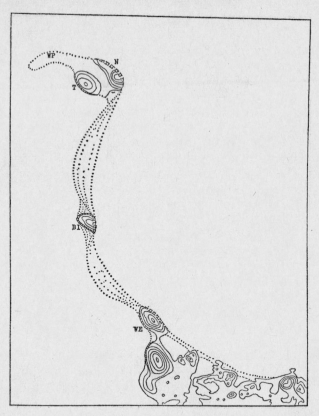

Fig. 7.8.

Great Hill gone, the beach would connect Nantasket Hill, Little
Hog Island, Strawberry Hill, White Head and Sagamore Head.
Strawberry Hill would be at an exposed angle of this beach and would
soon be destroyed. Little Hog Island and White Head would be
more exposed than before, providing the former had outlasted
Strawberry Hill. Sagamore Head and Hampton Hill would take their
turns in controlling the position of the beach until completely reduced

by the wave attack. The drumloidal extensions west of White Head and the Hull district at the north will be the last remnants of Nantasket Beach to survive, and of these two the Hull district will probably last much longer. If this interpretation is essentially correct, the relations in the Nantasket region will, in the remote future, resemble those indicated in Fig. 7.8. It is possible that the connecting bars may be broken through by the sea in one or more places, and that sand spits may replace the bars here shown. This will depend on local conditions of water depth and other factors which cannot be predicted. At present the area here shown is shallow, and favours the building of bars as indicated.

The protection of Great Hill is the key to the preservation of the entire Nantasket Beach district. A sea-wall has been constructed for the preservation of Little Hill, and this, of course, means protection to the adjacent areas of Great Hill. By such protective measures man may indefinitely postpone the normal changes which Nature would effect in the Nantasket area. It is interesting to note that man has begun his work in controlling the development of Nantasket Beach just at the time the beach has reached the greatest size which Nature could probably give it. Heretofore the beach has been increasing in area. Hereafter the normal development of the beach, unless arrested, would result in decreasing its area.

CHANGES OF LEVEL IN THE NANTASKET AREA

Much has been written concerning possible elevations and depressions of the Massachusetts coast since the glacial epoch. The evidence is often unsatisfactory and contradictory, but is thought by many to indicate a gradual subsidence at the rate of approximately 1 ft in one hundred years. Professor Crosby believes that some of the drumlins, which now show no marked cliffs facing towards the Atlantic, were strongly cliffed before Nantasket Beach was completed, and that subsidence has carried these cliffs under water. 'This view relieves us of the necessity of imagining a cordon of drumlins outside of the present beach which have been completely washed away, although it is not improbable that Harding's Ledge and the Black Rock Islets are the foundations of such vanished drumlins' (Crosby, 1893, p. 170). As we have shown above, there is abundant evidence that a number of drumlins did formerly exist outside of the present beach, and that these drumlins and their associated bars and spits effectively protected drumlins back of them from erosion. No subsidence is required to account for the lack of cliffing on the eastern ends of drumlins back of the present beach, and no evidence of submerged marine cliffs has ever been found.

It seems to us quite possible that there may have been a considerable depression in the Boston region since the glacial epoch; and that there may have been a very recent depression *of small amount* at the calculated rate of 1 ft in one hundred years. But that there has been any marked change in the relative position of land and sea during the last thousand years or more seems to us absolutely incompatible with the evidence furnished by Nantasket Beach. West Beach, as had already been pointed out, is similar in size and elevation to the beaches being formed along the present eastern shore of the Nantasket area. Had there been marked depression since the formation of West Beach, that beach would now be very low, possibly completely submerged. Had marked elevation occurred, West Beach should be relatively high, and other evidences of elevation should appear along the western margin of this beach. The close similarity between the oldest and latest beaches in the Nantasket area proves that the sea stood at about the same height when the two were formed. The intervening beaches are often low, because of the rapidity with which the shoreline was prograded for a time; but County Road Beach is strong and high, and may be compared with West Beach and the recent beaches.

The duration of this still-stand of the land may be roughly calculated. Judging from old maps, there has been no marked change in the width of Nantasket Beach during the last two hundred years. Judging from the rate of cliff cutting in various drumlins in the vicinity of Boston as determined by surveys extending over forty years or more, the length of time required for the removal of those portions of drumlins which have disappeared since the early cliffing of Strawberry Hill and White Head and the formation of West Beach, with liberal allowance for relatively rapid cutting of drumlins well exposed to the sea, could scarcely have been less than one thousand years, and was probably two or three thousand years. We conclude, therefore, that there have been no marked changes in the relative position of land and sea in the Nantasket area during the last thousand years at least.

CONCLUSION

The form of Nantasket Beach presents a variety of complicated phenomena which, when carefully studied, enable us to reconstruct with reasonable certainty the history of the development of the beach. It appears that the present form of the beach is not due to the accidental tying together of a few islands without system, but represents one stage in a long series of evolutionary changes which have occurred in orderly sequence and in accordance with definite physiographic laws. Perhaps nowhere in the world can features of beach development be better studied than in the area here under investigation. Certainly nowhere in the literature is recorded an example of so complicated a shoreline preserving the records of its past development with such fidelity.

REFERENCES

CITY OF BOSTON (1902–3) List of Maps, Appendix I, *Ann. Rep. City Eng.*, supplement.

CROSBY, W. O. (1893) 'Geology of the Boston Basin: Part I. Nantucket and Cohasset', *Bost. Soc. Nat. Hist. Occasional Papers*, no. 4.

DAVIS, W. M. (1896) 'The outline of Cape Cod', *Proc. Amer. Acad. Arts and Sci.*, XXXI 303–32.

GILBERT, G. K. (1883–4) 'The topographic features of lake shores', *U.S. Geol. Surv., 5th Ann. Rept.*, 69–123.

—— (1890) 'Lake Bonneville', *U.S. Geol. Surv., Mon. I.*

GULLIVER, F. P. (1899) 'Shoreline topography', *Proc. Amer. Acad. Arts and Sci.*, XXXIV 149–258.

JOHNSON, D. W. (1906) 'The New England Intercollegiate Geological Excursion, 1905: geology of the Nantasket area', *Science*, n.s., XXIII 155–6.

UNITED STATES COMMISSIONERS (1865) *Communications and Reports in relation to the Surveys of Boston Harbor, 1859–1865* (Boston), especially the *Second Report, 1860*, City Doc. No. 97.

8 Methods of Correlating Cultural Remains with Stages of Coastal Development

WILLIAM G. McINTIRE

INTRODUCTION

Students in the field of Recent geology have frequently overlooked the potentialities of using man's cultural remains as a tool to ferret out answers to problems of coastal development. Although it has long been recognised that cultural remains were indicators of subsidence in deltaic areas, few studies using cultural change have been conducted in coastal areas. In the 1930s Howe, Russell and Kniffen recognised the importance of using Indian remains as a tool in understanding the complex physical setting of coastal Louisiana (Howe *et al.*, 1935, pp. 64–8; Russell, 1936, pp. 162–70; Kniffen, 1936, pp. 407–26). Kniffen's reconnaissance survey indicated that there was a relationship between man's cultural change and the sequence of stream systems (Kniffen, 1936, p. 426).

Since 1951 Louisiana State University has conducted coastal studies for the Office of Naval Research, and these studies have provided an opportunity to test further the use of cultural remains in the Mississippi deltaic plain. The methods used and some examples of their application will be discussed in this paper to illustrate better the possibility of gaining important data by the correlation of cultural material with the physical setting. The 15,000-square-mile area of the Mississippi deltaic plain (Figs. 8.1, 8.2) is a natural laboratory for studying the processes of sedimentation, submergence, river changes, coastal growth and retreat and other related processes. In an area dominated by near-sea-level lakes, marshes, swamps, bayous and tidal channels, the prominent landmarks are natural levees, salt domes, cheniers[1] and beaches.

[1] 'Chenier' is used in south-western Louisiana to mean old beaches now stranded in marsh (Russell and Howe, 1935). 'Cheniere' is used in the southern part of the Mississippi Delta to mean any high ground and ordinarily refers to natural levees of abandoned channels. In both cases the name refers to the oaks, which are dominant among the trees covering such eminences (Russell, 1936, p. 45).

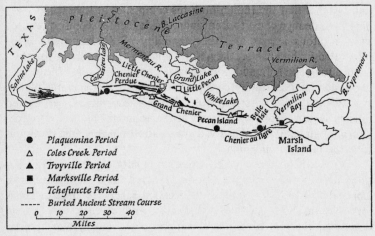

Fig. 8.1   Map: Distribution of initial occupation sites correlated with chenier sequence

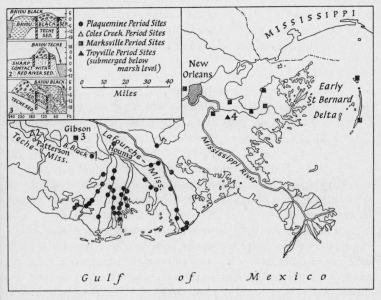

Fig. 8.2   Map: Distribution of initial occupation sites in St Bernard and Lafourche–Mississippi deltaic areas

William G. McIntire

It was on the sanctuary of the natural levees, salt domes, cheniers and beaches that the Indian built his camp sites at least 2,000 years before white men came to Louisiana. Aside from discomforts caused by ubiquitous insects and the possibility of his camp site being inundated by high water, the natural landscape provided a bountiful food supply in a humid, subtropical climatic setting. The waters abounded in fish, molluscs and numerous edible aquatic plants. Animals and fowl were plentiful, as were fruits, tubers, berries and nuts in season. Probably because of the abundant natural resources and mild climate, man has lived continuously in this area during at least the last two millennia.

During this time-period the processes of submergence, sedimentation and wave erosion have gradually altered the coastal features of Louisiana. The Mississippi river extended several distinct distributary systems into the Gulf of Mexico, and as each new system developed the loss of sediment supply in the old system allowed deterioration of the land mass and retreat of the shoreline. Each new system provided a new habitat for man, while deterioration of the old forced him to move. Whenever the requirements for human habitation were no longer available along a certain stream, man was forced to find other home sites, and thus there is a direct relationship between the places where he lived and the stages of deltaic development.

The physical evidence of Indian habitation in coastal Louisiana is found in numerous identifiable shell middens, black-earth middens and shell or earth mounds. Middens, which are the most common, represent the lived-on areas and are formless heaps of refuse intermixed with ash, bones, shell, dirt, pottery fragments and other debris of daily living. Mounds are hillocks intentionally formed by the Indian builder and were used for burial or ceremonial purposes.

Pottery fragments or potsherds found in mounds and midden accumulations are of special interest to the investigator correlating cultural remains with stages of coastal development. They were used in the Louisiana study primarily because they were the only universal cultural items that have withstood both time and the elements. In other areas different artefacts may provide a comparable measurement. Potsherd characteristics and designs were studied and through their differences a chronological series of cultural complexes was established for the Lower Mississippi Alluvial Valley (Phillips *et al.*, 1951). From the oldest to the youngest the major culture complexes are: Archaic, Tchefuncte, Marksville, Troyville, Coles Creek and

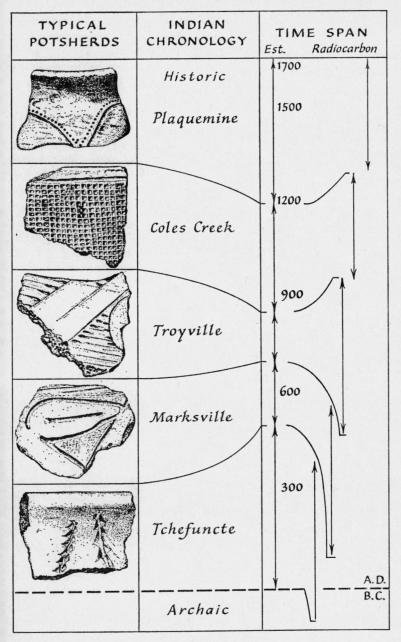

TYPICAL POTSHERDS	INDIAN CHRONOLOGY	TIME SPAN
		Est. Radiocarbon

Fig. 8.3   Culture complexes

Plaquemine periods (Fig. 8.3). Although the Archaic culture is included in the chronology, there has been no definite evidence located to indicate that Archaic people occupied the coastal area. They may have lived in the region, but the remains of their culture have apparently been lost because of subsidence below present sea or marsh level.

The other five periods are characterised by relatively distinct pottery types, as shown on Fig. 8.3. When potsherds recovered from mounds and middens are analysed and classified, the minimum age of the natural levee, chenier or beach where they were found can be determined. Within each site more than one Indian period may be present, but it is the oldest or initial occupation that is used to determine the minimum age of the chenier or levee upon which the camp site is located. Indian potsherds are also found along the entire coast of Louisiana, with the exception of the modern delta area. These wave-washed sherds were once located upon natural levees or beaches of more seaward-extending distributary streams, and have been washed up on the beach with the usual collection of light bulbs, beer cans and other drift debris.

In addition to human artefacts, types of shell in various combinations of stratification give important clues to changes in water salinity, sedimentation and submergence. From nature's abundant storehouse of foodstuff the Indian seemed to rely heavily on the brackish-water clam, *Rangia cuneata*. Both the *Unio*, a freshwater clam, and the *Ostrea* or oyster which requires saline water conditions, were also utilised by early man in coastal Louisiana. Although a few *Unio* shells are found in sites throughout the survey area, the greater part of the shell heaps are made up of *Rangia*. *Ostrea* shell is found in considerably smaller quantities than the *Rangia* and becomes even less frequent in inland sites.

There are several methods of applying and interpreting man's artefacts, his midden refuse heaps and artificial mounds. Indian remains were used in the Louisiana study, but other groups of people and their cultural remains could be studied and applied to physical problems in a like manner. The relationship between initial occupation sites and their physical base shows the stratigraphic sequence of stages of sedimentation. Many sites show layers of different types of shell which are indicators of environmental and salinity changes. Throughout coastal Louisiana Indian mounds and middens are tools for measuring active subsidence by local compaction of sediments as well as geosyclinal downwarping. The distribution of initial occupation sites on natural levees and cheniers, as well as surface-collected sherds from the beaches, shows stages of coastal development. Lastly, Indian remains are used to interpret deltaic deterioration. After the initial coastal study was completed, radiocarbon age dates were determined on samples of charcoal, bone and shell taken from key sites throughout the survey area, and in general the dates supported the previously estimated chronology for the sequence of delta and chenier development. The radiocarbon method provides absolute dates and enhances the potential uses of cultural remains.

STRATIGRAPHIC POSITION OF CULTURAL REMAINS

The stratigraphic position of human remains is not a new concept in the study of coastal landforms (Russell, 1936). In many places in the world the presence of cultural remains has helped the student to a better understanding of sequences of sedimentary processes. Cultural remains were useful in the investigation conducted in Louisiana because they substantiated some physical interpretations and further indicated that either a reinterpretation of some areas was necessary or that additional data were required.

Before the Indian story was worked out, it was established that the Teche–Mississippi system (Fig. 8.2) was an old course of the Mississippi river and that the Red River occupied the Teche channel for some time following the diversion of the master stream to its approximate present position. After the Red River abandoned the Teche channel, Bayou Black, which is a distributary of the Lafourche system, entered the channel but reversed its direction of flow for some miles before breaking through a natural levee and reaching the coastal marsh (Russell, 1940, p. 1202). The system of dual levees and overlapping identifiable river deposits makes a clear story of the sedimentary history. Indian site investigation not only supports the established chronology but also aids in delimiting the time at which the events occurred. Fig. 8.2 shows the location and cross-sections through three respective Indian sites associated with the Bayou Teche complex. Borings in the Gibson site show stratified Red River deposits intermixed with clam shells from which Marksville-period pottery fragments were recovered. The interpretation is that the Mississippi river had abandoned the Teche channel prior to Marksville time and that the Red River was then occupying the old Teche course. A Coles Creek site near Patterson is based on Red River

ICD G

deposits with no intermixing, and indications are that the Red River had abandoned the Teche course by Coles Creek time. A third site near Houma is of Plaquemine period and is associated with the Bayou Black sediments which blanket the older Red and Teche River deposits. These three sites show a direct relationship between stream sequence and Indian chronology, suggesting a minimum time when the Teche channel was occupied by the Teche–Mississippi, Red River and Bayou Black.

SHELL SEQUENCE AND SALINITY CHANGES

Molluscs were an important source of food to coastal Indians, and since it would have been difficult to carry them over great distances it is probable that their villages were close to the mollusc source. Many middens are extensive heaps of shell; some are several hundred feet long by 50 to 100 ft wide. Several of these sites show a shell sequence from the midden bottom to the top. When *Ostrea* and *Unio* are associated with *Rangia* they reflect a different physiographic environment, and in some sites stratification of different shell types occurs. Since the three types of mollusc require different water conditions, the presence of two or more types in one site (Fig. 8.4) indicates a change in water salinity. The range extends from freshwater *Unio* through brackish-water *Rangia* to the *Ostrea*, which requires more saline water. Shell sequences from the bottom to the top range from fresh to saline and saline to freshwater conditions, depending upon the sedimentary record and proximity to the Gulf.

Both shell stratification sequences are found in Indian middens in the western part of coastal Louisiana. Shell found in the Tchefuncte or oldest Indian middens associated with inland cheniers is dominantly *Ostrea*. One site on Little Chenier near the Mermentau river (Fig. 8.1) has a 1-ft base of *Ostrea* shell which is capped by about 2 ft of *Rangia*. In the middens of Marksville and younger periods the shell is primarily *Rangia*. During the time that the Tchefuncte Indians lived on Little Chenier ridge, it is possible that the coastline was either at the ridge or much nearer than the present shoreline, which is approximately six miles distant. In addition to the *Ostrea* shell found in the middens there are osyter reefs at shallow depths below the marsh just behind the Little Chenier ridge complex.

In the northern area of Grand Lake, at the confluence of Bayou Lacassine and the Mermentau river and near the mouth of Vermilion

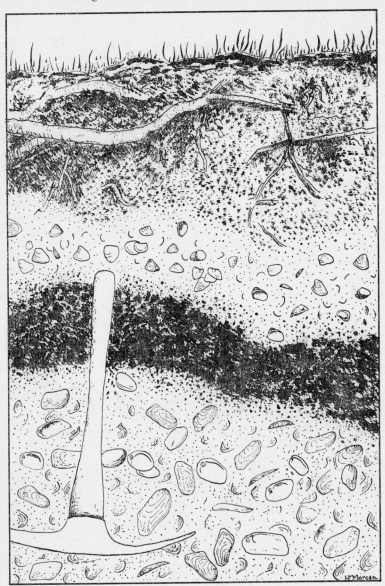

Fig. 8.4   Shell midden showing shell stratification. Unio *at the base and* Rangia *at the top separated by about ten inches of black earth (drawn from photograph of site).*

river (Fig. 8.1), some midden bases are *Unio* capped with a solid layer of *Rangia*. Several sites in the Atchafalaya Basin north of Patterson also have *Unio* bases overlain by *Rangia* (Fig. 8.4). The base of the midden in Fig. 8.4 is approximately 10 in. of *Unio* shell and levee material topped by 6 to 10 in. of black earth, which is capped with several feet of *Rangia* shell.

Similar examples of shell sequences are found in eastern coastal Louisiana; the predominant sequence pattern in this area, however, is from brackish clams at midden bases to oysters at the top. This shell stratification reflects the gradual deterioration of the area and the ensuing encroachment of the sea since the river ceased to flow eastwards. Different sequences of shells indicate changes that have occurred in water salinity and suggest changes in the sedimentary history of the area. When correlated with other techniques used by coastal geographers, different shell sequences in Indian mounds and middens become useful aids in understanding the stages of coastal development.

SITE SUBSIDENCE

Most low deltaic coastal areas are sinking, either by compaction of sediments or by geosynclinal downwarping (Russell, 1936, p. 162). Coastal Louisiana is a deltaic mass of unconsolidated material that has been subsiding throughout most of its history. The tremendous load of sediments deposited at the mouth of the Mississippi river every day is gradually depressing the area. The fastest rate of subsidence is recorded in the proximity of the river mouth, and gradually becomes less inland and on its flanks.

Indian mounds and middens have long been used as a tool for measuring subsidence, and throughout coastal Louisiana early man's camp sites have undergone varying degrees of submergence. Because they are completely submerged below present marsh level, there are probably many Indian sites that will never be found. However, many shell middens have been exposed along stream banks by erosion and excavated by canal dredging. A network of artificial waterways criss-crosses the coastal region: the Intracoastal Waterways span the state; numerous commercial companies have dredged their own canals; trappers, hunters and fishermen have altered the waterways for their own convenience and road-builders have dredged canals to obtain road-building material. Very often valuable information about the area that would otherwise have been much more difficult to obtain has been uncovered by canal dredging. In areas where known sites are not exposed by canals or streams, borings and/or excavations are necessary to determine the contents of the middens and amounts of submergence. The recovery of data on subsidence of Indian sites is primarily by laborious hand-auger borings, but the work is usually rewarding.

In several instances there was no evidence of a buried levee system on aerial photographs or on the ground, and dredged materials provided the only key. Pottery fragments were frequently present in the dredgings and from them a relative age could be determined. Borings in the submerged sites indicated the amount of subsidence, established that they were built on former natural levees of an active stream, and provided material about the processes of sedimentation and vegetative growth that had buried the levee–midden complex. From this recovery point the ancient levee system was traced to its origin by additional borings. In other areas the tops of mounds or middens protruding above the marsh level were the only indication of the existence of an ancient buried river system.

East of the Mississippi river, in the St Bernard deltaic area, an oil company dredged a canal through an Indian shell midden that had been completely buried below marsh level (Fig. 8.2, site 5). In the spoil bank were shells and pottery fragments which were classified as of the Troyville period. Borings through the site showed about 2 ft of marsh on top with the base of the midden 6 ft below marsh level. The borings also showed that the midden was sitting on an ancient natural levee. Beginning with this initial information provided by the Indian site, the ancient distributary was traced back towards its master stream. This is one of a number of examples where an Indian site has revealed subsidence, as well as keys for determining ancient river courses that have been masked by recent marsh growth and sedimentation. It is quite probable that the existence of this ancient distributary would never have been recognised but for the Indian material found in the dredgings of the canal.

Subsidence depths for the bases of shell middens range from near sea-level in the western Chenier area and in other areas where Pleistocene deposits are near the surface, to depths of 13 ft in the central and eastern part of the deltaic region. The greater depths are usually associated with older Indian sites, and shallow depths are related to later Indian periods. Borings through two Marksville-period sites in the St Bernard deltaic area showed subsidence amounts of about 11 ft below present marsh level, whereas borings in several Plaquemine-period sites on the Lafourche–Mississippi distributaries indicated subsidence depths of 1 to 2 ft. The original height of the natural levee is unknown, so by using present marsh level as a datum a minimum amount of subsidence can be ascertained.

SITE DISTRIBUTION AND BEACH COLLECTIONS

After all the field data were gathered in the Louisiana study, the pottery was classified according to the chronology and mapped aerially on the basis of the initial occupation of the site. When the sites were mapped, a pattern was revealed which showed a definite relationship between different time periods of man's occupance and the habitable area in which he settled. The distribution of initial occupation sites on natural levees and cheniers, and sherds collected from the beaches are important keys to river changes, coastal retreats and time sequence. Pottery distribution can also be used to show deltaic deterioration and its relationship to offshore beaches or marsh flats.

Initial occupation on distributary systems

The distribution of initial occupation sites in the St Bernard deltaic area (Fig. 8.2) was one of the first clues that the region may have been much older than was previously thought. The courses of most of the ancient streams that were flowing eastward in the St Bernard area are masked by more recent sediments or have submerged to near or below marsh level, and therefore cannot be mapped from aerial photographs or traced out on the ground surface. Borings into Marksville sites, which are the oldest sites recognised in the area, showed that they were based on ancient natural levees that generally trended eastwards. The indications are that prior to Marksville time a major Mississippi distributary system flowed into the area and extended its ancient delta beyond the present Chandeleur Island arc. Surface-collected Marksville pottery along the beaches of the Chandeleur Islands suggests its former connection with the deltaic mass. When the master-stream sedimentation was cut off and the coastline retreated into the old delta mass, remains from Indian sites on former streams were added to the Chandeleur beach matrix. The presence and distribution of ancient Marksville sites in the St Bernard region led to a re-evaluation of the Mississippi river chronology.

Unlike the St Bernard area, which is older and has undergone an extensive amount of deterioration, the distributaries of the Lafourche–Mississippi are relatively well preserved and the relationship between site and stream is more easily discernible. The oldest pottery found in sites associated with this distributary system is of Plaquemine age (Fig. 8.2). It was formerly believed that the Lafourche–Mississippi system was older than the St Bernard system, but distribution of initial occupation sites in both areas showed the opposite sequence of sedimentation. The youngness of the Lafourche and the antiquity of the St Bernard deltas were established first by the initial occupation method and were later substantiated by radiocarbon dating.

Initial occupation on cheniers

In the chenier–plain complex of western Louisiana the distribution of initial occupation Indian sites shows a relative sequence of development with major chenier ridges (Fig. 8.1). When the initial occupation sites on individual cheniers were mapped, the distribution showed a relationship between initial Indian occupation and chenier development. The oldest sites (Tchefuncte) are on the most northern ridges and the youngest (Plaquemine) on the beaches and most seaward ridges. Between the younger and the older ridges are two cheniers that show relative age difference. Chenier Perdue was occupied first by Troyville-period Indians, and the front ridge of Grand Chenier was first settled by Coles Creek-period people. The distribution of the oldest sites on the northern ridges and progressively younger sites on Chenier Perdue, Grand Chenier and Chenier au Tigre indicates a direct relationship between initial occupation and stages of development.

Furthermore, when the initial occupation on cheniers and stream systems are correlated, a relationship emerges between delta and chenier development. When the master stream was in proximity to the chenier region, mud flats were deposited in front of the cheniers and the shoreline prograded. After the sediment supply was cut off by the river shifting its course eastwards the shoreline retreated and new cheniers were formed. This relationship is shown by the distribution of Plaquemine-period sites on the Lafourche–Mississippi distributaries (Fig. 8.2) and the presence of Coles Creek pottery on Grand Chenier and the Front Ridge of Pecan Island (Fig. 8.1). The Grand Chenier complex was formed during the time the master stream was flowing east of the Lafourche–Mississippi; when the river changed its course to the Lafourche system, the extensive mud flats between the Grand Chenier complex and Chenier au Tigre were deposited. The sediment supply was again cut off when the Mississippi abandoned the Lafourche system for its present position and Chenier au Tigre was formed. The presence of Plaquemine-period pottery on Chenier au Tigre completes the latest phase of the relationship between chenier–delta development and Indian cultural remains.

Surface-collected potsherds from beaches

With the exception of the modern delta area, which is too recent for Indian habitation, the remaining coastline of Louisiana is a fairly good hunting ground for Indian potsherds. Sherds were collected from the entire shoreline, and although many were badly wave-washed, identification was often possible. The major portion of the shoreline is retreating and the presence of potsherds indicates that at one time the distal section of the streams extended farther sea-wards. As the encroaching sea gradually destroyed a deltaic area from which sedimentation had ceased, sherds from Indian middens were washed ashore and remained as part of the beach matrix. When beach-collected potsherds were classified chronologically and the distribution of the oldest sherds was mapped, the pattern showed a chronological sequence for various segments of the shoreline (Fig. 8.5).

In general, surface-collected beach sherds and sherds collected from stream sites in a related area substantiate one another (Fig. 8.5). In the oldest deltaic areas, however, extreme subsidence and deterioration has obliterated the original Indian record along the coasts. This is particularly true of the area west of the Lafourche–Mississippi system, where only late-period sherds are present. In the Cheniere Ronquille area beach-collected sherds of Troyville period suggested an older delta land mass than the adjacent stream-site sherds indicated. Borings in the marsh behind Cheniere Ronquille showed the natural levees of an ancient buried stream; the Troyville sherds collected from the beach were probably from the Indian site associated with it. In this case beach-collected sherds provided the clue, and with additional borings the presence of the older buried system was established. The sherds from the beach were the same type as those collected from Bayou Barataria and its distributaries and are probably related to the sites on the same river system.

More recent shoreline studies (Morgan and Larimore, 1957) show that retreat rates are related to stages of delta development.

Retreat rates are greater along the younger coastal segments and are progressively lesser along the older delta fronts. There is a close parallel between delta-front retreat rates and the age indicated by potsherds collected from the beaches.

Indian evidence for deltaic deterioration

A cursory study of topographic maps of coastal Louisiana between the western chenier marginal plain and the Chandeleur Islands

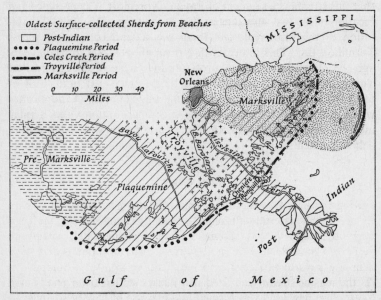

Fig. 8.5 Map: Deltaic areas of initial occupation and oldest surface-collected sherds from beaches

shows identifiable coastal sections which are related to known major delta areas. Aside from the modern delta, which is building seawards, the remainder of the coast is retreating and segments of the delta fronts are in various stages of decay. Through subsidence, cessation of sedimentation and wave attack the amount and degree of deterioration reflects the relative ages of delta masses. The younger is well defined and in a better state of preservation, whereas the older is less distinct, with little or none of the original delta surface remaining above gulf level. The Vermilion Bay area (Fig. 8.1), which is in an advanced stage of decay, represents one of the oldest identifiable delta areas. Bayou Cypremort, which was a major stream

ICD H

in this area, is now truncated by the retreating northern shoreline of Vermilion Bay. At the mouth of Bayou Cypremort is a shell heap of wave-washed *Rangia* shell and pottery fragments of Tchefuncte period; across Vermilion Bay, west of Marsh Island, a Marksville shell midden whose surface is at marsh level is exposed along the present retrograding shoreline. Borings through the midden indicate that it was associated with an ancient natural levee system whose stream scar is barely discernible on aerial photographs. It is thought that this stream is a segment of Bayou Cypremort. The presence and location of the two sites suggest that at the time of the initial inhabitation of the Marksville site there was a land connection across Vermilion Bay. This indicates that from about 1,500 to 2,000 years ago the old delta mass was probably in a stage of decay and that Vermilion Bay has progressively enlarged since that time. East of the Vermilion Bay area there are numerous examples of Indian sites which are clues to stages of deltaic deterioration, and similar methods would apply in relating Indian occupation with deteriorating deltaic segments.

SUMMARY

During the past two thousand years that man has lived in coastal Louisiana his material culture has gradually changed from one form to another, and the measurement of these changes delineates time periods. The cultural remains found primarily in mounds and middens are keys to processes and sequences of coastal development. As the master stream and its distributaries shifted from one course to another, new land was formed and because of lack of sedimentation the old land gradually deteriorated. Whenever lack of human requirements forced man to leave his old home sites, he moved to areas where they were available. Concentrations of initial occupation sites of a particular time period in one area and the absence of them in another shows regions that were habitable during certain stages of delta development.

Although sedimentation, water salinity, subsidence, deterioration and stages of coastal growth and retreat are all indicated by cultural remains, one of the most important relationships is between initial occupation sites and stages of chenier and stream development. When initial occupation sites are mapped for an entire area, many interrelated correlations between streams, cheniers and beaches and the sites based upon them are readily apparent.

This method produces clues for interpreting the physical area as a whole, as well as providing more detailed information about segments of a given area. Total mapping of initial occupation sites presents a pattern that is subject to field checks and is useful in setting the stage for additional detailed geological research.

REFERENCES

FISK, H. N. (1944) *Geological Investigation of the Alluvial Valley of Lower Mississippi River* (Mississippi River Commission, Vicksburg, Miss.) 78 pp. and plates.

HOWE, H. V., RUSSELL, R. J., and McGUIRT, J. H. (1935) 'Geology of Cameron and Vermilion Parishes', *State of Louisiana, Dept. of Conservation*, Geol. Bull. No. 6, 242 pp.

KNIFFEN, F. B. (1936) 'A preliminary report on the mounds and middens of Plaquemines and St Bernard Parishes, Lower Mississippi River Delta', *Dept. of Conservation, Louisiana Geol. Survey*, Bull. No. 8, pp. 407–22.

McINTIRE, W. G. (1958) *Prehistoric Indian Settlements of the Changing Mississippi River Delta* (Louisiana State University Press) 127 pp. and plates.

MORGAN, J. P., and LARIMORE, P. B. (1957) 'Changes in the Louisiana shoreline', *Trans. Gulf Coast Assoc. Geol. Soc.*, VII 303–10.

PHILLIPS, P., FORD, J. A., and GRIFFIN, J. B. (1951) *Archaeological Survey in the Lower Mississippi Alluvial Valley, 1940–47* (Papers Peabody Mus. Amer. Archaeol. and Ethnol., Harvard Univ.) XXV.

RUSSELL, R. J. (1936) 'Lower Mississippi River Delta: reports on the geology of Plaquemines and St Bernard Parishes', *Dept. of Conservation, Louisiana Geol. Survey*, Bull. No. 8, pp. 1–193.

—— (1940) 'Quaternary history of Louisiana', *Bull. G.S.A.*, LI 1199–1234.

—— and HOWE, H. V. (1935) 'Cheniers of south-western Louisiana', *Geog. Rev.*, XXIV 449–61.

9 Coastal Research and its Economic Justification

PER BRUUN

ABSTRACT

Proper and thorough planning of coastal engineering projects is discussed and the economic justification of research work indicated. Examples are given concerning navigational problems, coastal protection problems and harbour sediment problems.

THIS paper is written as a causerie. No attempt has been made to base its reasonings and conclusions on a dollar-and-cents basis, but rather it stresses the importance of common sense, good science, good technology and – most important – good conscience. 'All that you do – do with all your might. Anything done half is never done right.'

'We have no time for that sort of thing and furthermore we have no confidence in it' has been the standard excuse for lack of proper and thorough planning of many coastal engineering projects, whether they comprised a navigation problem, a coastal protection problem or a harbour sediment problem. The result was in one case a continuous struggle to keep an inlet free from deposits as a result of inadequate dredging – the use of inadequate equipment at inadequate time intervals. Another result was inadequate coastal protection planning – taking chances in some respects and over-dimensioning in other respects, thereby leaving the arena to engineering philosophy instead of to engineering science.

The question of *why we do coastal research* is therefore not difficult to answer: It is necessary to know and understand the coastal phenomena in order that we can:

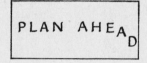

because if we do not do just that the result we come up with may look as foolish as the above figure.

Coastal research includes a great number of subjects, ranging from the emplacement of huge breakwaters on the ocean bottom for the purpose of checking ocean waves and sand drift, to the planting of proper vegetation in marsh areas and on dunes for checking sediment transport by water or wind. The employees involved in this research programme are recruited from a great variety of fields in arts and science: geology, geography, soil mechanics, coastal engineering, hydraulic engineering, oceanography, physics, mathematics and meteorology. In order that a coastal set-up shall be a complete and fully effective organisation, it must include people from all of these fields, who in mutal good understanding 'carry the ball' of coastal research.

A discussion of the economic justification of such research requires a discussion of the applied sides of research aspects, but it should never be forgotten that without fundamental research applied research of any importance is impossible.

Man's interference with coastal development is most powerfully manifested in 'the jetty' – the huge monsters jutting far out into the ocean as 'artificial promontories' built either as vertical impermeable monolithic block jetties or as sloping rubble-mound or block jetties which are permeable for water to some extent but will not allow the passage through them of wave motion or littoral drift material.

Jetties are not a new invention. They were built thousands of years ago. The ancient port at Alexandria with the famous lighthouse, Pharos, had rubble-mound jetties. Jetties for the ancient harbour at Tyre which were discovered recently have massive stone breakwaters which in construction showed a notable advance over the work at Pharos, as there were two walls of hewn stone, keyed together with metal dowels, the space between the walls being filled with some kind of concrete. The Greek harbour jetties were founded upon beds of tipped impervious material, with masonry forming the super-structures. Roman harbour jetties, as found, for example, at Ostia, were much more substantial than anything previously existing both in design and in construction. One reason was the Roman cement, which contributed to stability and lasting qualities. Methods of constructing underwater works were involved and all Roman jetties and breakwaters were built of masonry founded at sea-bed level. A structural technique – based on experience – was already highly developed.

Medieval ports using monolithic rock or rubble-mound design and mostly having a wharf on the protected side were built in Italy,

Great Britain, France, Holland and Germany. Realising through costly experience the forces hidden in wave action, harbours were almost without exception established in protected estuaries, bays and waterways.

With the rapid development of navigation in the nineteenth century, it became necessary to construct harbour jetties out into the open sea in countries such as France, Great Britain, Italy and Spain. The design varied from place to place, but the desire for saving materials usually resulted in attempts being made to build the jetties with as steep slopes as possible – frequently as block constructions founded on rubble-mound layers on the bottom. Such jetties were often subject to extremely strong wave action and heavy damage occurred. It is no wonder that attempts at a rational approach to the design of jetties based on wave forces started in these countries.

The first research concentrated on measuring wave forces in the prototype (England, Italy and France), and it became clear that there is a wide difference between wave forces exerted by deep water, shallow water and breaking waves – the latter giving rise to extremely high shock pressures of explosive character thanks to an air pocket often associated with the breaking phenomena.

A mathematical approach to the problem of wave forces by trocoidal waves on a vertical wall was presented in 1928 by Sainflou, whose theory was later tested by Cagli in full-scale measurements of wave action at Genoa (Italy), and by Rouville and Petry at Dieppe (France). Based on these full-scale tests, diagrams were developed which proved useful for practical design.

Meanwhile research is still not satisfactory for the shallow area where trocoidal characteristics are changed to solitary and breaking wave characteristics, and where, moveover, the direct influence of wind cannot be neglected. Most harbour jetties are located in this particular area of changing and irregular wave characteristics, and most research in prototype and in hydraulic models (Bagnold, Cagli and the U.S. Corps of Engineers) has therefore been concentrated on forces by waves which were breaking or about to break. There is considerable scatter in the results of these tests, model tests indicating comparatively much higher shock pressures than prototype tests with irregular wave trains (Dieppe, France). The collapse of, or heavy damage to, extremely expensive breakwaters such as those at Antofagasta (Chile), Catania (Algiers), Alderney (England) and Bilbao (Spain) could probably have been avoided if wave mechanics had been explored beforehand and certain precautions taken against too heavy forces by breaking waves. Other jetties, e.g. the Dover Admiralty Pier, have stood even the heaviest wave action. It is indeed surprising that research in this particular and economically well-justified field so far has hardly involved a total expense exceeding the cost of a hundred metres' length of one of the collapsed heavy-duty breakwaters. Much research work is awaiting proper action; the U.S.S.R. recently announced comprehensive research in fact on waves and wave forces to be carried out from a breakwater where rooms for research equipment were built into the breakwater itself.

Regarding the detailed design, costly experience has shown that attempts in only 'estimating' the proper size of blocks (whether natural or artificial) and other pertinent factors for the stability of the jetty are often quite costly, and difficulties gradually developed in regard to meeting the costs of numerous mishaps. Because of the complexity of this problem all rational approaches must be considered as semi-theoretical, inasmuch as experience coefficients play an important role in their composition (Iribarren, Kaplan, Hudson, Hedar, U.S. Waterways Experiment Station).

Let us, from a purely engineering, wave mechanics and structural field, move into coastal morphology, founded by the geographers (Davis, 1912; Johnson, 1919; von Richthofen) and utilised later by the engineers who needed its results in order to understand and predict the natural development of certain coastal areas for which harbours or other coastal developments were planned. The importance of coastal research in practical life is clearly demonstrated in the harbours built at Dublin, Ireland (Fig. 9.1). Unsuccessful attempts at maintaining desirable depths in the estuary of the Liffey river led to the construction of the harbour at Howth and later to the development of Dun Laoghaire harbour further south. The physical situation is that flood currents with the normal 10-ft tidal range run north while ebb currents run south. Both currents make turns into the bay part of the river entrance. Prevailing winds are from the south and west, but the biggest waves enter the area from the east.

The bay area is greatly bothered by deposits of river and littoral drift sediments. With the construction of the harbour at Howth less trouble was expected. Meanwhile, it was unfortunate that elementary principles of coastal morphology and littoral drift technology were not considered, and the heavy sand drift from the north along the concave shoreline towards the north-west caused large deposits along the western jetty, eventually covering it completely. The third attempt at establishing a harbour was the construction of the harbour at Dun Laoghaire. Because of this location littoral drift materials from the north do not penetrate into the Dun Laoghaire area, and even if the flood currents from the south-east carry considerable amounts of solids, these materials are not deposited in the harbour entrance, partly because of its advantageous configuration and partly because the slow outgoing ebb currents in the entrance are able to hinder penetration of materials into the harbour itself.

These harbours were all built in the nineteenth century, when the field of coastal morphology was in its infancy. It is therefore unfair to blame the design engineers for their mistakes, which nevertheless

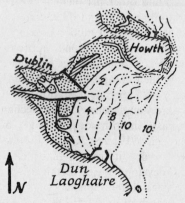

Fig. 9.1 Harbours at Dublin, Ireland

were of a rather elementary nature. Meanwhile similar mistakes have been made in the twentieth century, e.g. in Italy where uncritical use of the Italian engineer Cornaglia's 'neutral depth' theory for sand transport towards or away from the shore led to a number of great failures, e.g. Maurizio Harbour. The Danish version of the same theory, the so-called 'Headland-theory' (*Pyntteori*) also led to a couple of rather expensive and not very successful experiments at Hirtshals and Hanstholm, which are North Sea coast headlands. Attempts are now being made to correct these mistakes.

Proper research in, and knowledge about, coastal morphology could have decreased the amount of trouble and saved taxpayers the cost of expensive corrective measures.

Coastal morphology takes into consideration not only the development of the coast in plan (Bruun, 1954; Davis, 1912; Ferguson, 1959; Schou, 1945; Steers, 1945), but the development of beach and bottom profiles as well (Bruun, 1954; Saville, 1950). In order to evaluate the stability of a beach and its 'foundation', the offshore bottom, knowledge about their reactions to wave and current activity is necessary. These problems have been studied for years by coastal researchers. It is now known that beach and bottom profiles are subject to seasonal fluctuations depending upon the change in wave action from one time to another. It is also realised that their slopes cannot develop beyond a certain maximum steepness, but on the other hand it has been shown that they are 'tough-stable' and do not collapse suddenly like a piece of structural engineering, e.g. a bridge or a piece of aerodynamic engineering such as an aeroplane (as was claimed in Denmark by a coastal committee in 1942 regarding the stability of the Thyborøn barriers. The claim resulted in inadequately planned protection on one side and the taking of unnecessary risks on the other side. Further unnecessary precautions were taken by over-dimensioning other elements, such as the time factor.) A 'glass of cold research ice water' would have permitted a more thorough and better justified plan from the very beginning. This is now all realised and is being corrected. The author of this article is only concerned that, whatsoever is suggested, it shall be well thought out, well reasoned and tested technically as well as economically.

Speaking about sedimentation, the simplest problems are those in rivers and canals, which should be mentioned briefly because of their relation to coastal problems. It is no wonder that important developments within this field were the result of research work in India, the United States and the U.S.S.R., where enormous flood, irrigation and drainage problems call for proper planning and, therefore, the assistance of research. China, however, is the country which has experienced the great flood disasters. Millions of Chinese have through the years lost their lives in floods caused by inadequate river regulation and drainage, brought about not least by uncontrolled sedimentation in rivers. The Chinese have now become very active in this research, where basic knowledge of physics and mathematics is so important, and this fits the Chinese mind.

British engineers in India made the first contributions to the practical sedimentation technology, introducing the so-called 'regimen theories' as a basis for design of drainage canals (Sir Claude Inglis). Engineers in India and Thomas Blench in the United States later followed up this line, while the U.S.S.R. and Germany took more interest in the physical aspect of channel stability. Meyer-Peter's work in Switzerland, Shield's work in Germany and Kalinske's work in the U.S.A. further developed this field, which entered its purely physical and final development stages in the work of Einstein (1950) and Chien in the U.S.A. The results were better planning, fewer mishaps and, therefore, huge savings. One of the bad examples of planning which ignores sedimentation laws was the construction of a huge hydraulic power plant in the Congo. Shortly after its completion the plant became choked up with sediment and the project had to be re-worked (by model experiments).

Let us from this introduction return to sediment problems on seashores, where they are mainly concentrated around harbour and coastal protection works.

Sedimentation problems on seashores and their relation to man-made structures can most effectively and conveniently be explained by the terminologies 'source' and 'drain'.

A *source of materials* is a coastal zone, submerged or emerged, which delivers materials to other coastal areas. A source might be an area where erosion takes place, a shoal in the sea located for instance on the downdrift side of a (newly) jetty-improved inlet, the shallow area in front of an inlet which has been closed, a river which transports sand material to the coastal zone, or sand drift from dunes to the beach. Artificial nourishment of any kind to a beach is also a source.

A *drain of materials* is a coastal zone where materials are deposited. Natural drains include marine forelands of any kind such as spits, recurved spits, tombolos, cuspate forelands, angular forelands, etc. A drain may also be a bay, an inlet or a shoal. Artifical drains include man-made constructions such as jetties, groynes, dredged sand traps, inadequately designed and inadequately constructed harbours, etc.

In practical coastal engineering and littoral drift technology the following rules are valid:

1. A coastal protection should be built in such a way that it functions as a drain. It should therefore have a source but not a drain on the updrift side. If there is a drain the coastal protection in question cannot be expected to work satisfactorily unless materials are supplied artifically to the shore in question.

2. A harbour (or an improved inlet) on a littoral drift coast should not act as a drain. It is therefore desirable that it has no source area or only a limited source area on the updrift side or on

either side of it. It is best if it has a drain on the updrift side or on both sides.

Without making themselves fully clear on the importance of 'sources' and 'drains', geographers, geologists and engineers have, with great eagerness, studied these phenomena for decades, the geographers concentrating on the coastal morphology aspects (Johnson, 1919; Schou, 1945; Steers, 1945), the geologists most often on the mineralogical aspects, and the engineers on the total amount of nuisance caused by inadequate understanding and therefore lack of respect for nature's source and drain rules and regulations (Abecasis, 1955; Bruun, 1954; Cornick, 1959; Schijf, 1959).

Let us consider a few of these cases. Miami Beach, Florida, is provided with a great number of wooden or steel groynes. There is, however, very little beach left and statistics indicate that only approximately 15 per cent of the visitors to this famous beach and seaside resort ever swim in the ocean. The 85 per cent prefer to stay on the dry side of the shoreline or else enjoy swimming in the numerous swimming pools which have now been built. The natural conclusion seems to be that Miami Beach is not a very attractive beach for ocean bathing, and the reasons for this apparently are a scanty and not very attractive beach, steep offshore bottom, dangerous currents at the vertical wall groynes, and too much loose shell material (up to 50 per cent) in the beach sand.

Using coastal engineering terminologies, the reasons could also be expressed as a result of the lack of any source of material for the groyne system in question. It has probably cost several millions of dollars to build up coastal protection at Miami Beach, mainly based on groynes and vertical sea walls, and the result is that little beach is left. If a source of suitable material for beach nourishment had been located in the bay and this material had been dumped on the beach, we would still have had, and could still maintain, a beach at Miami Beach instead of great amounts of coastal protection junk.

Another example is Palm Beach, Florida. After the inlet was dredged, the jetties which were built in 1918–25 had blocked the southward littoral drift almost completely. The consequence was heavy erosion on the south side of the inlet. Over a number of years attempts were made to combat this erosion by the construction of a great number of groynes, but being without any source of material the groynes failed. Modern development in the coastal protection field was later responsible for artificial nourishment from the bay and finally (1958) a by-passing sand plant was put in operation on the north side of the inlet and is supposed to pump 200,000–250,000 cu. yds of sand fill across the inlet per year. Farther south it is the intention to nourish the beach from dredging operations in the bay. It would probably have been better if groynes had never been built.

It is a well-known fact that groups of groynes function as drains, and for this reason will always have adverse effects on the downdrift shore. If groynes were not drains they would not work at all (Schijf, 1959). It may nevertheless not be fully recognised that groynes will usually cause considerably more erosion than accretion! A good example of such tremendous disadvantage to the overall picture is seen in a group of 130–250-m.-long groynes on the Danish North Sea coast at Bovbjerg. The groynes in question have stabilised the beach where they were built, but on the lee side (south side) they have caused erosion of the shoreline of up to 10 m. per year in farmland. It is now the intention to build more groynes on the 2-km. non-protected downdrift shore extending to the next group of groynes, which consists of only five partly abandoned shorter struc-tures. Meanwhile the result will only be an extension and activa-tion of the erosion problem farther south.

This again points with adequate clearness to the fact that artificial nourishment of beaches is to be much preferred as shore protection because it is entirely free of skirmishing 'boundary conditions'. Meanwhile, in order to utilise artificial nourishment it will be neces-sary to develop better and more suitable dredging equipment, such as nuclear-powered submarine dredges as suggested by the author in an article in *Shore and Beach* (American Shore and Beach Preservation Association) in June 1959.

Harbours are not supposed to work as drains for littoral drift materials. They are supposed to work contrarily. They can, however, be built in such a way that they present marvellous 'Olympic gold-medal drains' through not being designed correctly. The harbour of Madras, India (Fig. 9.2), presents a very instructive case (Cornick, 1959). Its breakwaters extend outwards about 1,000 m. from the original low-water shoreline (1876). Up to 1913 a large triangular area of sand, about 260 acres (105 hectares) in extent, had accumu-lated on the south side of the harbour; on the north side considerable shoreline recession had taken place. The old entrance to the harbour was centrally situated between the breakwaters facing east, and the sand drifting northwards found slack water between the pier heads in which to settle, with the result that before the entrance was closed it was shallowing up at the rate of about 1 ft per year. In 1902 a north-east entrance project was started, including a 400-m.-long sheltering arm completed in 1911. The result of this closing of the old entrance and extension of the eastern arm caused continued deposition along the whole eastern jetty face, and this would have become more and more pronounced if it had not been checked by comprehensive dredging operations. Another advantageous result of the described 'remodelling' was that the harbour became smooth enough for working cargo into and out of lighters alongside the ships and piers in practically all kinds of weather. Later another sheltering arm was built at the southern corner of the harbour, where

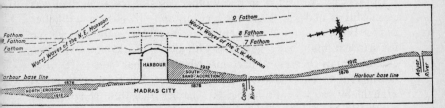

Fig. 9.2   The Harbour of Madras, India (H. F. Cornick)

accumulating sand is checked by a suction-dredge (mounted on the arm) which pumps the spoil into hopper barges moored inside the harbour. All the expensive nuisance described above could have been avoided with proper planning, but the hydraulic model technique was unknown when the harbour was first built between 1870 and 1880.

The same is true for Zeebrugge harbour in Belgium, which was completed in 1907. It has the configuration of a big northward-curved 'nail'. Heavily silt-laden flood currents, running at 5–6 ft per second (1·5–2 m.) from the south-west carried 3–4 million cu. m. of silt per year into the harbour to be deposited on the lee side of the jetty in a big eddy current. Attempts were made to flush this material away by a 400-m.-wide *clair-voie* (opening) at the land end of this jetty, but the result was an increase rather than a decrease in the deposits. The opening was therefore closed, and after the Second World War hydraulic model experiments were carried out, partly in Holland and partly in Belgium, to solve this problem. By constructing a large semicircular breakwater to fill out the eddy area, deposits in the harbour will decrease by about 50 per cent, which in turn will present a tremendous saving in maintenance costs of the harbour. The remainder of the material by-passes the harbour with the tidal currents.

In somewhat similar model experiments with the Karlsruhe river harbour in Germany, special jetty configuration secured the by-passing of heavy bed-load transport in the river flow.

The harbour at Abidjan, Ivory Coast, Africa, presents a similar problem which was properly solved by model experiments in the Netherlands. A cut was made to connect the ocean with a lagoon to accommodate vessels of 27 ft draught (Fig. 9.3). Sand coming from the west is deposited by the flood current at M; the ebb current, which is strongly concentrated at that point, transports it in the direction of P, where part of it settles in a deep hole in the sea bottom.

In this case, as well as in many other cases of research, man was successful in making nature his servant, and this is so much better than making nature an opponent or enemy. This philosophy is true for artificial 'man-made' harbours with jetties, breakwaters,wharves, etc., as well as for natural harbours which man has tried to improve in different ways. This last-mentioned subject has been given much thought by coastal morphologists, whether they were geographers, geologists or engineers, and deserves special mention because of its relationship to one of the most interesting subjects in coastal research.

The ancient Egyptian, Phoenician, Greek, Roman and Viking naval fleets were based in estuaries, bays, fjords and lagoons, and we find similar installations today at such places. Now, as thousands of years ago, the tidal estuary, river or inlet is a cultural factor of immense importance.

It is customary to talk about 'nature's delicate balance' which man cannot touch without bringing about adverse effects. The fact is that everything in nature is in a process of development and man, by interfering with this development, can influence the natural process in one way or another; the accompanying effects will usually be adverse in certain ways but advantageous in other respects.

Inlets have always been 'problem children', and this is particularly true of those inlets which have resulted from breakthroughs on littoral drift shores – and this includes the greater part of them (Brown, 1928).

Lack of understanding of inlet physics led to misuse of inlets, particularly when they were loaded with more navigation responsibility than they were able to carry. The result was an endless succession of failures. There is hardly an inlet on the United States barrier east coast, or the Danish North Sea coast, which has not caused all kinds of trouble, including irregular shoaling or deepening, uncontrollable meandering, erosion or accretion, unprovoked movements, or even sudden 'disappearances', not to mention headaches, backaches and ulcer trouble. This is true whether the name of the inlet is Ponce De Leon, Great Egg, Man-Killer (Matanzas) or Thyborøn. The reason why they were 'problems' was because they were not 'understood', and for a long time their various 'doctors' were representatives from all branches of life, including butchers and

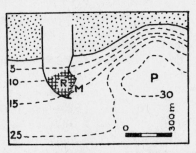

Fig. 9.3   Abidjan Harbour, Ivory Coast, Africa.

lawyers, but not coastal researchers who were able to handle the problem from a physical point of view. Brown (1928) and O'Brien (1931) were responsible for the first real progress, which was later followed up by the work of others on an entirely physical basis (Bretting, 1959; Bruun and Gerritsen, 1958, 1959; U.S. Army Corps of Engineers, 1959; Dronkers and Schönfeld, 1955). It is now known that an inlet in alluvial material is not only a 'difficult hole' but that it – as everything else in nature – depicts a balance between the acting forces. Based on analysis of many inlets (Bruun and Gerritsen, 1958) it seems possible to express the stability of an inlet 'Stab' as:

$$\text{Stab} = F\left(ts, \frac{O}{M}, \frac{Qm}{M}\right)$$

where *ts* is the so-called 'stability shear stress' between flow and bottom. ($ts = \dfrac{Pg^{v2}}{C_2}$, where P = density of water, g = acceleration of gravity, V = mean velocity of flow, and C = Chezy's friction coefficient.) O = the so-called 'tidal prism', which is the total amount of water flowing through the inlet in one half tidal cycle, usually referring to spring tide, flood or ebb conditions, and M = the amount of littoral drift material brought to the inlet entrance per year. (For Qm, see below.)

Consider first *ts*. A great number of analyses of inlets have shown that the cross-sectional area of the inlet, i.e. the smallest cross-section in the inlet channel, can be considered explicitly as a function of different factors such as maximum flow, configuration and shape of the cross-section flow characteristics, shear stress between flow and bottom, soil conditions, suspended load and littoral characteristics, wave action, freshwater head flow and finally the 'time history' of the inlet channel. These factors are interrelated and further analyses have shown that the shear stress *t* is the most practical and useful parameter (Bruun and Gerritsen, 1958). The question of inlet stability has therefore become a 'structural design problem' in which detailed computations of flow (Dronkers and Schönfeld, 1955) must be compared with 'the allowable' or 'the ultimate strength' of the bottom ('the determining shear stress', *ts*), which in turn depends upon the factors mentioned above. For 'average conditions', *ts* is about 0·39 kg. per sq. m.; for heavy littoral drift conditions, 0·47 kg. per sq. m.; and for light littoral drift conditions, about 0·32 kg. per sq. m. It is hopeless to endeavour to maintain an inlet with free flow over an alluvial material bottom with lesser values of *ts*.

Meanwhile satisfactory $\dfrac{O}{M}$ and $\dfrac{Qm}{M}$ ratios are as important as an adequate *ts*. Consideration of a great many inlets (Bruun and Gerritsen, 1958) has revealed that those having an $\dfrac{O}{M}$ ratio in excess of 300 have a higher degree of stability, while inlets with $\dfrac{O}{M}$ ratios < 100 have a more predominant transfer of sand on (shallow) bars across the inlet and less significant tidal currents, for which reason they are rather unstable and usually characterised by narrow, frequently shifting channel(s) through shoals. It is not possible to say where the transition $\dfrac{O}{M}$ ratio between stable and unstable inlet channels lies because the littoral drift irregularity, in quantity as well as in direction, will most likely make it impossible to establish such a fixed ratio. Meanwhile numerous mishaps could have been avoided if such rather elementary problems had been investigated and taken into consideration properly before actual construction work commenced. In fact, regardless of where you go in the world, the philosophy seems to have been that 'everybody shall have his private inlet exactly where he (not nature) pleases' (University of Florida, 1958).

The question of an adequate $\frac{O}{M}$ ratio automatically brings to light the fact that littoral drift material – even with the most advantageous ts and $\frac{O}{M}$ – cannot pile up infinitely on either side of an inlet's seashore or on sea and/or bay shoals. It is necessary to get rid of this material by passing the material across the inlet channel either by natural or artificial means.

If nature itself in numerous cases did not by-pass sand across inlets, passes and channels on seashores, a number of 'marine forelands', including barriers, spits and entire peninsulas, would not exist. A typical example of nature's strategy is found in Florida, which was built up of sand washed down by rivers and streams from the Appalachian Highland and carried southwards, crossing estuaries and tidal inlets, for final deposition in the huge barrier and ridge systems which we call Florida. In fact northern Florida seems to be the world's largest recurved spit system (Schou, 1945; University of Florida, 1958).

The two main principles in the by-passing of sand by natural action are:

By-passing on an offshore bar, and
By-passing by tidal flow action.

Most cases present a combination of these two methods.

A submerged bar in front of an inlet or harbour entrance on a littoral drift coast will often function as a 'bridge' upon which sand material is carried across the inlet or entrance (Bruun and Gerritsen, 1959). Every channel dredged through the bar will therefore be subject to deposits.

By-passing by tidal flow action takes place when littoral deposits are spilled out of the inlet by ebb currents in the downdrift direction. Both bar and tidal flow by-passing include cases of irregular transfer of large amounts of materials in migrating sand humps or by change in the location of channels.

Research (Bruun and Gerritsen, 1959) has revealed that one can distinguish between inlets or entrances with predominant bar by-passing and inlets with predominant tidal-flow by-passing by considering the ratio $\frac{M}{Qm} = r$ between the magnitude of littoral drift (M in cu. yds per year) and the quantity of flow through the inlet or entrance (Qm in cu. yds per second under spring-tide conditions).

If this ratio is >200–300, bar by-passing is predominant; a ratio <10–20 indicates that conditions for predominant tidal-flow by-passing exist. Meanwhile whether or not such by-passing actually takes place depends on whether or not it is possible to use the tidal flow for transferring material in the downdrift direction. This depends, among other things, upon the inlet configuration. Inlets exist which, owing to strong tidal currents, expel material so far out into the ocean that it is lost for ever to the shore. Characteristic examples of this situation are Fort Pierce Inlet and Bakers Haulover Inlet in Florida, where inlet ebb currents up to 7–8 ft per second may occur, particularly at the Haulover Inlet (University of Florida, 1958). Similar current velocities may exist in Thyborøn channel after a storm, when the tide is running out and shooting material out into the North Sea.

By-passing problems can be solved by careful planning, including model experiments such as were carried out for the harbours at Abidjan, Lagos, the Volta river and many others. Failures and heavy maintenance costs have in this way been avoided. Establishment of sand traps, including devices for artificial (mechanical) by-passing, is an example of man's 'cut-through' of the problems when other solutions were not convincing or possible, as for instance at Palm Beach Inlet, Florida.

The sediment transport field is still in a state of rapid development, the radioactive tracing technique being the newest invention. Two different types of radioactive labelling are now in use: the direct labelling and the artificial labelling. The direct labelling can be realised either by neutron activation of sediment constituents (as with the phosphorus P^{32} St Peter quartz sandstone from Kentucky) or by absorption into or the depositing on the sediment's surface of a radioisotope, such as radioactive gold Au^{198} (used in California), and radioactive silver Ag^{110} (used in Portugal). The artificial labelling is employed by the solution of a radioisotope in melted glass, which when ground and properly screened is supposed to reproduce the properties of the sediment. The best tracers seem to be the isotope scandium Sc^{46}, which has been used in rivers (the Thames) as well as in the sea (off the Norfolk coast). The Russian luminophore method uses fluorescing materials.

The Sc-tracing technique was developed particularly in Great Britain (Dept of Scientific and Industrial Research, 1957, 1958; Inglis and Allen, 1957). An example of the use of Sc^{46} is the now classic river Thames experiment carried out in 1954 and 1955 by the

Hydraulics Research Establishment, Wallingford. The isotope Sc[46] was selected as a suitable gamma-ray source, with a convenient half-life of 85 days. The Thames experiment was arranged with the object of demonstrating with certainty whether or not landward transport of silt takes place in the Thames estuary. The tracer material had a density similar to that of Thames mud, and consisted of soda glass containing about 1·5 per cent of scandium oxide. Material corresponding to 30 curies was injected in the main shipping channel abreast of the entrance to the tidal basin of Tilbury Docks at the upper end of Gravesend Reach, 26 miles below London Bridge. No dredging of the shoal area at the lower end of Gravesend Reach during the period of 18 days immediately preceding injection was carried out, so that the radioactive material would not be unduly attracted there. Immediately prior to the time of the test a systematic blank survey was made of background readings on Geiger counters on the bed of the estuary between 8 and 38 miles below London Bridge. The scandium glass was mixed with natural mud and released from containers on the river bottom, after which detection started. One of the surprising results obtained during the next three weeks of tracing was that in the tidal basin at Tilbury Docks (12 miles above the injection point), where siltation necessitates considerable dredging, the activity gradually increased to three times the background value during the first fortnight. From the total number of observations it became quite clear that silt can move towards the head of the estuary in these reaches when it is known that close to the bed there is a net landward movement of water. This in turn indicates that dredged material should be pumped ashore behind the high-water line. This change of practice, compared to the present dumping in the outer part of the estuary, would not be expected to have an immediate effect on the river because regime is a delicate balance between accretion and erosion, and as material was removed it would be partly replaced by material eroded from the mud flats and by fine silt from the coast washed into the estuary on flood tides. Some of the silt would be deposited in the estuary instead of being washed seawards on the ebb as hitherto. Gradually, however, the balance would change until eventually a considerable improvement would occur, with a corresponding reduction in the amount of dredging required. The economic importance of this would be enormous.

Similar techniques are now under development for the seashore, the U.S.S.R., Great Britain and Portugal having the lead so far.

It is indeed much better to have nature as your friend than as your enemy.

Typical examples of a somewhat different method of making nature a real enemy are presented in the numerous vertical coast-protection sea walls built everywhere in the world, whether they are heavy gravity walls of the English type or steel sheet-pilings such as the Florida shores are cluttered with – many of which are turned over or are in other ways worn out because of misunderstood use

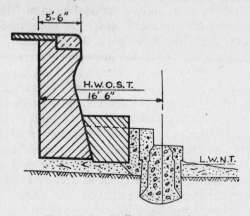

Fig. 9.4   Sea Wall at Bray, Ireland

and inadequate design (University of Florida, 1958). Fig. 9.4 shows a gravity wall at Bray, Ireland. It was built in 1884–6 with cross-section as shown by heavy full lines. Its vertical face contributed to an increase of erosion at the same time as oversplashed water and inadequate drainage aggravated its stability. It therefore became necessary to put a sheet-piling apron in front of the wall, but its vertical face had the same adverse effect as the original wall. Finally it was necessary to put one more (caisson) apron in front of the former apron, and all of this became very expensive. Today Florida continues the same mistakes made in Ireland seventy years ago. The collapsed sea wall of Jacksonville Beach in Florida clearly demonstrates what happens when an equal amount of misunderstanding of the problem and lack of proper planning made up the prevailing background for the design. Some miles of similar sea wall collapsed in Florida, and more will collapse in the near future because Florida has been due for a serious hurricane flood for several years now.

The statistical approach to storm flood-tide analysis was 'born' in Holland. In 1939 Wemelsfelder published a statistical analysis of high-tide data on the Dutch coast. His method, when adjusted to and interpreted in agreement with the local situation, allows estimation of the frequency of high tides and also, using great care, extrapolation outside the zone of present experience. Such frequency analyses, now in progress in Florida and elsewhere where storm tides are common, are of great importance, e.g. for the determination of the insurance values of real estate in coastal areas. In Florida, despite the lack of adequate data, the available information clearly shows that the possibilities of flooding are high and, unfortunately, very much underestimated. At many coastal communities and developments even the most elementary considerations with respect to safety of life and damage to property have been disregarded and the inhabitants are living on 'borrowed time'. Those who 'developed' the coastal areas in question are not easy to find, but many occasionally appear behind the so-called 'free press' when they believe that this will help them force their dollar-based desires and inadequate projects through.

The above examples all consider 'wet parts' of the coastal research fields. Other parts are only half wet or perhaps all dry. A company built a rubble-mound breakwater pier somewhere in the United States. This pier was supposed to carry pipelines for fuel oil. It was indicated that a conservatively rigged pipeline would not be a proper solution under the given circumstances, since it was built in that 'headed-for-trouble-way' and became an expensive 'baby' for this reason. A little research – such as suggested – would have saved the company tens of thousands of good American dollars.

Half of wet coastal work includes reclamation of land in swampy areas and in marshland. Here again it is true that the intelligent method of procedure is to let selected plants do reclamation work instead of hauling in all the dirt over perhaps long distances. Examples of such reclamation work are found in the British and Dutch *Spartina* grass marshland and in the Danish reclamation work on the North Sea coast.

The dry counterpart to this vegetation reclamation are the measures against sand drift by proper plants such as *Ammophila* species (helme). Where formerly wind blew away dunes and piled up sand on roads and agricultural land, proper plantings have been able to build up dunes and dykes where they were wanted for coastal protection reasons, such as the Danish west-coast sand dykes. In the United States, Cape Hatteras National Park is now using mechanical planting machines pulled by crawler-type tractors and

developed by local research. The practical dune-planting research
by the National Park Service is expected to decrease the unit price
of planting to about 50 per cent of the cost of conservative methods
of planting by hand.

Plants have been imported to Florida from North Carolina and
Denmark and are doing fine, but more research is necessary to find
plants which will fit the different clamatological zones.

Fig. 9.5   Layout of the Delta Project, Netherlands (Van Veen)

Let me finish this 'sermon' on coastal research by mentioning one
of the largest – if not the largest – coastal engineering research
projects the world has ever seen, which is the research programme
associated with the Dutch Delta Project. This huge undertaking was
initiated after the 1953 flood disaster, which killed approximately
2,000 people and caused a billion dollars' worth of damage.

The setting of the Delta Project is shown in Fig. 9.5 (Ferguson,
1959). It includes three big dams in the river entrances and two
smaller ones farther inland. The waters of the Delta area will be
divided into two separate basins by means of dams. The southern
basin will be entirely cut off from the sea and become a freshwater

lake. The northern one, which comprises the mouths of the Rhine and Meuse rivers, will continue to be connected with the sea because the waterway to Rotterdam must remain open to shipping. Tidal waves will therefore still be able to penetrate inland by way of this basin, but they will only cause high tides in the waterway itself.

In order to secure the best and most economical result from this huge project, the cost of which may be as high as one billion dollars before it is completed, in approximately twenty-five years, the Dutch have undertaken an extensive research programme including research on tides, tidal currents and density currents in the Delta area itself and in the connecting areas. Furthermore, detailed studies of wave action and sand movement are under way using the most modern techniques, including the establishment of permanent automatically operated 'pole-stations' out in the North Sea, which are loaded with instruments such as wind recorders, tide recorders, wave recorders, current recorders, etc.

Perhaps the most interesting part of the programme is the tidal research, including the influence of structures on the penetration of tides, whether they are of astronomic type or are mainly storm tides (Dronkers and Schönfeld, 1955). In the Netherlands no less than three different methods of tidal prediction are now in use: the hydraulic model, the computation method, and the electric analogue method. Each method has its typical merits and limitations. For some purposes one may be more suitable than the other. Perhaps a coastal researcher will get the most impressive view he can ever have by visiting Professor Thijsse's Dutch Nordoostpolder 'Open Air Laboratory', where up to thirty models from the Netherlands and elsewhere may be seen at one time.

If you ask the Dutch if all this research pays, they will most likely answer that 'they simply cannot afford not to do it'. Furthermore, you should remember that the Lord made the world but the Dutch built Holland'.

CONCLUSION

From the above causerie of examples of economic justification for coastal research it may appear that the author of this paper is inclined to believe that coastal research is something which we should always do, considering it at least from a face-saving point of view.

This is not the idea at all. He honestly considers it as being entirely irresponsible and foolish not to *plan ahead*, because nobody can afford to spend £25 million for a second-class product if he can secure a first-class product for £20 million or perhaps £30 million with the additional £5 million well spent for urgently needed improvements.

A designer's 'sense of responsibility' should always be related to knowledge about the safety factor, which he should have studied carefully before proceeding; and not to over-dimensioning of conditions and design on account of lack of adequate knowledge of the problem.

REFERENCES

ABECASIS, C. K. (1955) 'The history of a tidal lagoon inlet and its improvements (the case of Aveiro, Portugal)', *Coastal Engineering*, V.

BRETTING, A. E. (1958) 'Stable channels', *Acta Polytechnica, Scandinavia*, 245.

BROWN, E. I. (1928) 'Inlets on sandy coasts', *Proc. Amer. Soc. Civ. Eng.*, LIV.

BRUUN, PER (1954) *Coast Stability* (Copenhagen).

—— and GERRITSEN, F. (1958) 'Stability of coastal inlets', *Proc. Amer. Soc. Civ. Eng.*, LXXXIV, no. WW3, and *Proc. VIIth Internat. Conf. on Coastal Engineering*, in *Coastal Engineering*, VII.

—— and —— (1959) 'Natural by-passing of sand at coastal inlets', *Proc. Amer. Soc. Civ. Eng.*, LXXXV, no. WW5.

CORNICK, H. F. (1959) *Dock and Harbour Engineering*.

DAVIS, W. M. (1912) *Die beschreibende Erklärung der Landformen* (Berlin-Leipzig).

DEPT OF SCIENTIFIC AND INDUSTRIAL RESEARCH (1957) *Hydraulic Research* (Wallingford, Berks, England).

—— (1958) *Hydraulic Research* (Wallingford, Berks, England).

DRONKERS, J. J., and SCHÖNFELD, J. C. (1955) 'Tidal computations in shallow waters', *Proc. Amer. Soc. Civ. Eng.*, LXXXI.

EINSTEIN, H. A. (1950) 'The bed-load function for sediment transportation in open channel flows', *U.S. Dept of Agriculture, Technical Bulletin*, No. 10260.

FERGUSON, H. A. (1959) 'Hydraulic investigations for the Delta Project', *Proc. Amer. Soc. Civ. Eng.*, LXXXV, no. WW1.

INGLIS, SIR CLAUDE, and ALLEN, F. M. (1957) 'The regimen of the Thames estuary as affected by currents, salinity and river flow', *Proc. Inst. Civ. Eng.*, VII.

JAKOBSEN, B., and JENSEN, K. M. (1956) 'Undersøgelser vedrørende landvindings-metoder i Det danske Vadehav', *Geografisk Tidsskrift*, LV, and *Meddelelser fra Skalling-Laboratoriet*, XV (Copenhagen).

JOHNSON, D. W. (1919) *Shore Processes and Shoreline Development* (New York).

JOHNSON, J. W. (1953) 'Sand transport by littoral currents', *Proc. Vth Hydraulic Conference*.

—— (1951–8) 'Coastal engineering', nos. 1, 2, 3, 4, 5, 6, *Proc. Coastal Engineering Conferences* (Berkeley, Calif.).

LANE, E. W. (1955) 'Design of stable channels', *Proc. Amer. Soc. Civ. Eng.*, CXX.

NIELSEN, N. (1960) 'The organisation of scientific research work in south-west Jutland', *Geografisk Tidsskrift*, LIX.

O'BRIEN, M. P. (1931) 'Estuary tidal prisms related to entrance areas', *Civil Engineering*.

SAVILLE, T., JR (1950) 'Model study of sand transport along an infinitely long, straight beach', *Trans. Amer. Geophys. Union*, XXXI.

SCHIJF, J. B. (1959) 'Generalities of coastal protection', *Proc. Amer. Soc. Civ. Eng.*, LXXXV, no. WW1.

SCHOU, A. (1945) *Det marine Forland* (The Marine Foreland) (Copenhagen).

SHEPARD, F. P., and INMAN, D. C. (1951) 'Sand movement on the shallow inter-canyon shelf at La Jolla, California', *Beach Erosion Board, Technical Memorandum* no. 26.

STEERS, J. A. (1969) *The Coastline of England and Wales.*

UNIVERSITY OF FLORIDA (1958) *Selected Papers from Proc. Sixth Conference on Coastal Engineering.*

U.S. ARMY CORPS OF ENGINEERS (1959) *Bibliography on Tidal Hydraulics,* Committee on Tidal Hydraulics, Report No. 2.

VEEN, J. VAN (1936) *Onderzoekingen in de Hoofden* (Ministerie van Waterstaat, 's-Gravenhage).

—— (1948) *Dredge, Drain, Reclaim: The Art of a Nation* (The Hague).

—— (1950) 'Eb- en vloedschaar systemen in de Nederlandse getijwateren', *Tijdschrift Koninklijk Nederlandsch Aardrijkskundig Genootschap* (Amsterdam).

Index

Names of authors mentioned in papers and listed in the References are not given in the Index.

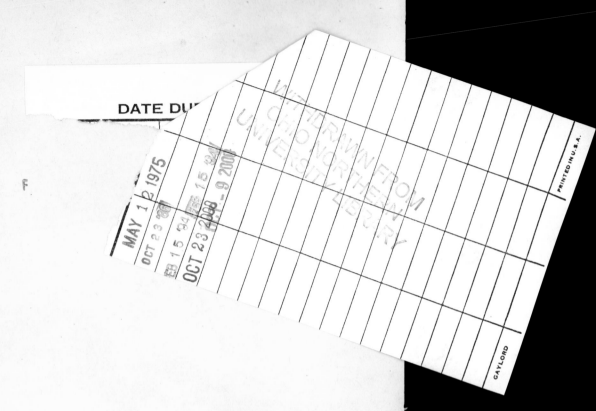